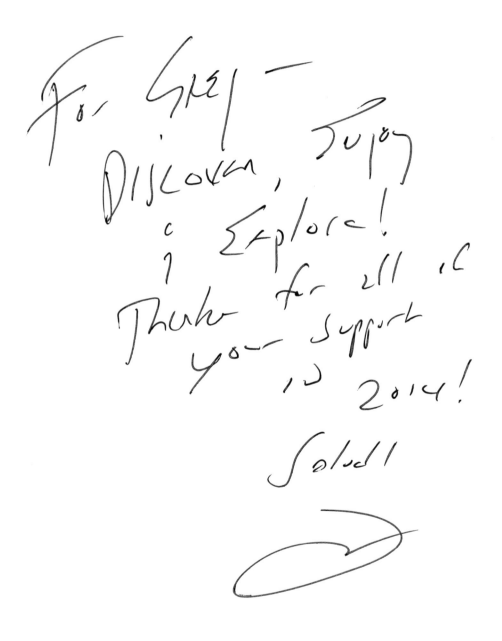

For Greg —
Discover, Enjoy
& Explore!
Thanks for all of
your support
in 2014!

Salud!

WINES OF SOUTH AMERICA

The Essential Guide

Evan Goldstein

UNIVERSITY OF CALIFORNIA PRESS

University of California Press, one of the most distinguished university presses in the United States, enriches lives around the world by advancing scholarship in the humanities, social sciences, and natural sciences. Its activities are supported by the UC Press Foundation and by philanthropic contributions from individuals and institutions. For more information, visit www.ucpress.edu.

University of California Press
Oakland, California

Library of Congress Cataloging-in-Publication Data
Goldstein, Evan.
    Wines of South America : the essential guide / Evan Goldstein.
        pages   cm
    Includes bibliographical references and index.
    ISBN 978-0-520-27393-1 (cloth : alk. paper) — ISBN 978-0-520-95875-3 (e-book)
    1. Wine and wine making—South America.   2. Viticulture—South America.   3. Wine industry—South America.   I. Title.
    TP559.S63g65   2014
    663'.20968—dc23
                                        2013047383

Manufactured in the United States of America

23   22   21   20   19   18   17   16   15   14
10   9   8   7   6   5   4   3   2   1

The paper used in this publication meets the minimum requirements of ANSI/NISO Z39.48–1992 (R 2002) (Permanence of Paper).

CONTENTS

MAPS

ACKNOWLEDGMENTS

The scope and diversity of those who contributed to make this book happen mirror the complexity and range of South America's wine regions. Reference books, by definition, require researching vast amounts of information, ensuring its accuracy, and tracking down minutiae. All of this requires asking favors of your friends, who in turn ask favors of their friends. So *muchisimas gracias* and *obrigadissimo* to

Barbara Pino, Elena Goldstein, and Adam Goldstein for their love, support, and empathy when I headed south of the equator for weeks on end

Connie Shih Cohn, my invaluable right hand, for her patience, persistence, and constant refusal to take no for an answer when it would have been far easier to throw in the towel

My business partner, Limeng Stroh, and our Full Circle team for rejigging their day jobs to allow me to take on this project

The amazing *equipo* at the University of California Press, including Blake Edgar and Dore Brown, for taking a chance on this book and then treating it so lovingly

Erika Búky for her fastidious editing skill, which still preserved my voice

Anne Canright, proofreader nonpareil, for dotting the i's, crossing the t's, and adding the accents

Lohnes + Wright for the captivating cartography

All of the gifted photographers, including Mariano Herrera, Laura Mariategui, Camilo Quintana, Lilo Methfessel, Mario Vingerhoets, and especially Matt Wilson, "the man," whose beautiful shots bring the countries to life

Jaime Draeger for her awesome transcription skills and great sense of humor

I am also grateful for the invaluable assistance of

Team Argentina: Raquel Correa, Laura Catena, Nora Favelukes, Mario Giordano, Victor Honoré, María Laura Ortiz, Magdalena Pesce, Andres Rosberg, and José Saballe

Team Brazil: Leocir Bottega, Gabriela Mott, Josimar Pedron, Pablo Onzi Perini, Flavio Pizzato, Bárbara Ruppel, João Carlos Taffarel, Jorge Tonietto, and Mauro Zanus

Team Chile: Fred Dexheimer, Ricardo Grellet, Derek Mossman Knapp, Aurelio Montes, Jake Pippin, Allyson Silva, and Lori Tieszen

Team Uruguay: Nicolas Bonino, Fabiana Bracco, Francisco Carrau, Pablo Fallabrino, and Gustavo Magariños

And to

Embedded emissaries Gregg Smith and Soledad Marroquin Muñoz (Peru), Sergio Prudencio (Bolivia), Dick Handal and Pablo Tamarelli (Ecuador), Gabriela Fines Ibarra (Paraguay), Jean Louis de Bedout (Colombia), and Leo d'Addazio (Venezuela)

Finally, I extend my gratitude to the countless winemakers and the helpful and endlessly generous teams at all the *bodegas* and *adegas* I visited for sharing stories, meals, laughter, and, of course, their delicious wines.

INTRODUCTION

The Vinos *and* Vinhas *of South America*

According to an old South American story, when God finished creating the earth, the angels in charge of shaping the land came to him and said: "We've got a lot of mountains, valleys, and rivers left over; what should we do with them?" God answered, "Dump them at the end of the earth." And that, fortuitously for grape growers and winemakers, is how Chile and Argentina were created.

My first encounter with South American wine was a revelation. On an October day in the mid-1980s, while I was working at my family's San Francisco restaurant, Square One, a friendly and passionate wine importer named Alfredo Bartholomaus came in. He was the founder of Billington Imports (then an independent company), and he had brought a few bottles of Chilean wine made by Cousiño Macul, one of Chile's oldest family wine companies. Having never tasted any South American wines, I stopped by his table and listened with a combination of curiosity and reticence as Alfredo shared his enthusiasm for Chile and its wines and invited me to sample the range he had brought with him. I was mightily impressed, if not totally surprised, and the tasting opened my eyes (and palate) to South American wines. I have never looked back.

I had been to South America once in the early eighties, but only to visit Brazil's big cities and beaches. On that trip, I hadn't tasted any local licensed beverages except beer and *cachaça* (a spirit distilled from sugarcane juice). Wine was not popular or readily available. Even today, beer (from every country), rum (from Venezuela or Brazil), and brandy (notably the celebrated piscos of Chile and Peru and the *singani* of Bolivia) are more popular than wine on most of the continent. Only Argentina and Uruguay are

1

serious wine-drinking nations. This lack of an indigenous wine culture is part of the reason that winemakers began to look overseas for their market, starting a veritable northward invasion of South American wines. And when I say invasion, I mean the good kind, like the British musical invasion of the 1960s—evocative and a great source of pleasure.

Today in many parts of the United States, a casual stroll down the wine aisles of a local store or supermarket reveals an array of wines from Chile and Argentina. Argentina is now the hottest South American wine producer and the third largest exporter to the United States, trailing only Italy and Australia. According to the Foreign Agricultural Service (FAS) of the United States Department of Agriculture, Argentina's 2013 bottled wine exports to the United States amounted to 7.2 million cases, worth $286 million, and grew a respectable 5.5 percent over the previous year.

Fifth among wine exporters to the United States, and only recently surpassed by Argentina, is Chile. According to the International Organisation of Vine and Wine (OIV), Chile is the world's ninth-largest producer of wine. It exports 879 million liters a year, totaling $1.88 billion in 2013 and accounting for 70 percent of the country's wine production. The United States is the second-largest market for Chilean wines after Europe, importing about $201 million worth.

In addition, vibrant wine industries exist in Brazil and in Uruguay, whose exports are increasingly becoming available to American consumers. Add to the mix some small but significant wine undertakings in Bolivia and Peru and a few in Colombia, Ecuador, Paraguay, and Venezuela, and you have a comprehensive, wide-ranging continental wine culture.

The arrival of these wines in the U.S. marketplace has been recent. My early encounter with Cousiño Macul's wines was the start of a pioneering effort. Before then, the export markets were only a minor focus of South America's producers. Today, political and economic changes have made some of the South American countries into global economic forces. In the wine industry, international investment and the advent of flying winemakers, as well as the development of local talent, have made South American producers a force to be reckoned with.

And yet, apart from a few sources of information on the wines, grapes, and vines of individual countries, there has been no up-to-date, comprehensive reference work on South American wine. Until now. I decided to write this book to fill this niche and to introduce more of the world's appreciative wine drinkers to the pleasures of South American wines. The October 2013 issue of *Decanter* magazine, with a focus on South America, observed that "no region has progressed so far or as fast in the past 10 years," and I concur.

We start out with an introduction to South America, take a look at the arrival of the vine and wine culture, review the range of grapes planted and cultivated, and summarize the developments leading to the current approaches in viticulture and winemaking. Separate chapters are devoted to the two wine-producing giants of the continent,

Argentina and Chile, examining their unique histories, wine regions, wine styles, and outstanding grape varieties, with leading producers listed for the various regions. The evolving wine industries of Brazil and Uruguay receive similar treatment. Another chapter summarizes current winemaking in Bolivia, Colombia, Ecuador, Paraguay, Peru, and Venezuela. Each chapter includes maps, statistics, and profiles of key individuals in the wine industry; it ends with a listing of notable wineries. Then we step back to a broader view once more, taking a look at food and wine culture in South America, particularly in Argentina, Chile, Brazil, and Uruguay. Another chapter offers practical advice for touring the wine regions of South America and their numerous *bodegas* and *adegas,* the respective Spanish and Portuguese terms for wineries. Finally, I offer a collection of "top ten" (or sometimes twenty) lists of recommended wines in specific and quirky categories and some suggested reading for South American wine pilgrims.

I'm a tour guide with opinions. On several visits to South America, I have done the heavy lifting and have selected from among the thousand-plus South American wineries those that I feel are consistent, noteworthy, or particularly promising. But in this book I do not play the role of wine critic. I don't give scores or ranks for wine. If a wine or a winery is included in this book, it is because I believe it to be contextually representative and meriting recognition.

I approached this project with some reservations. Until now, I have tried to write books that, like a fine wine, will age gracefully. Writing a wine guide is different. The wine world does not stand still: laws are modified, winemakers move on, wineries are sold or closed, great vintages sell out. For someone who loves precision, this reality is frustrating, because it makes complete accuracy impossible. In my previous books, I never discussed vintages (what do you do when you can't get the 2009 anymore?) or ephemeral details like annual sales and production volumes. This time, I have had to let go and acknowledge that some of the things I write about are moving targets. I hope you can accept, as I have done, the fact that the perpetual changes in the world of wine are what make it endlessly interesting as well as occasionally challenging to grapple with.

This book is intended to be an essential work, but it is by no means the last, definitive word on South American wine. So if there's a bottle you love or a winery you adore that isn't mentioned in the pages that follow, don't rise up in frustration; instead read on, discover more, and enjoy. *¡Salud!*

Map 1. South American wine-producing countries

OVERVIEW

If some higher being were to tell me my years with Old World wine
had to end, it would be to South America that I would turn.

Steven Spurrier

Vineyards in front of the White Mountain
(Cerro Blanco) in the Zonda Valley, San Juan,
Argentina. Photo by Laura Mariategui.

By any metric, South America is a significant contributor to the global wine market.
Although only the world's fourth-largest
continent, South America is the second
most important wine-producing continent,
after Europe. It embodies a number of geographic and climatic extremes, including
the world's highest waterfall (Angel Falls
in Venezuela), the largest river by volume (the Amazon), the longest mountain
range (the Andes), and the driest place on
earth (the Atacama Desert in Chile). This
diverse continent is also home to an amazing range of quality wines. A December
2012 Gallup poll cited seven of the world's
ten most upbeat countries as being in Latin
America, and wine surely plays a role in
creating that outlook!

For Northern Hemisphere wine lovers, South America challenges some common
assumptions. For one thing, southern seasons are reversed, with the autumn and grape

harvests arriving in the first half of the calendar year. In the north, we think of grapes typically being grown in a geographic band between 30 and 50 degrees latitude. Within this range, the climate offers sufficient overall warmth for wine grapes to ripen properly. At higher latitudes, the climate is too cold to ripen grapes reliably, and at lower latitudes (closer to the equator) the climate is too warm and often too humid. In the Southern Hemisphere, the same general principle applies. Here most grapes are cultivated between 23 and 45 degrees latitude. But in South America, viticulture extends far up on the slopes and highlands of the vast Andes mountain range, where the cooling effects of high altitude and the more intense solar radiation create unique growing conditions. And nearer the equator, some growers are taking advantage of the year-round warm climate to experiment with vines that produce harvests twice a year and in some locations, even three.

THE HISTORY OF WINE IN SOUTH AMERICA

Wine grapes are not native to South America, but they arrived as an essential adjunct to European settlement. After Christopher Columbus explored the northern coastline of South America in 1498, word of his discovery spread. From the early sixteenth century onward, European voyages and their accompanying migrations of European settlers led to the substantial exchange of animals and plants between the two continents. It was the Spanish conquistadores, who arrived in South America by way of Mexico, that gave birth to South American wine culture.

Although there is anecdotal evidence of indigenous grapes in Aztec civilization, the documented cultivation of wine in South America began during the time of the conquistadores. Led by Francisco Pizarro, they arrived in the sixteenth century, intent on finding treasure and conquering the Inca Empire, whose territory encompassed modern Peru, Bolivia, northern Argentina, Chile, and Ecuador. Between 1531 and 1534 the Incas were overmatched by Spanish weapons and military tactics, and the victors settled in. As they colonized the region, they planted vines from the seed that they brought with them, from grape varieties that we refer to today as criollas. They depended, however, on local technology: it was the irrigation engineering of the Inca emperor Pachacutec, the builder of Machu Picchu, that enabled the first vineyards to be planted south of Lima in 1548. As the conquistadores ventured farther south and east, grape growing and winemaking spread into Chile and northern Argentina.

Pedro de Valdivia, one of Pizarro's most trusted officers, settled Santiago (in 1541), La Serena (1544), and Concepción (1550). As he moved south, he was met with brave resistance by the native Mapuche, who ultimately killed him before he could cross the Bío-Bío River. He left a legacy of large-scale viticulture. By the late sixteenth century, Chile had widespread vineyards, producing wines from criolla varieties such as Pais (Criolla Chica), Muscatel (Muscat), Torontel (also known as Torrontés Riojano), Albihio (Albilla), and Mollar. Viticulture continued to expand in the sixteenth and seventeenth centuries.

By the time Chile declared independence in 1810, wine was culturally ingrained, especially around the silver mines of La Serena, where the miners were notorious for their insatiable thirst.

Early Spanish missionaries also cultivated grapes. A Jesuit priest, Francisco de Carabantes, is credited with introducing the first grapes, likely Pais, to the region that is now Chile in 1548. In 1556, Juan Cedrón, a Spanish missionary, made the trek from La Serena over the Andes to Santiago del Estero in northern Argentina, carrying cotton seed, grain, and vine cuttings, and so became Argentina's first winemaker. As in Chile, the first grape planted was likely Pais, possibly along with other criolla varieties. Adapting Inca irrigation methods, the local Huarpe tribes had harnessed the Mendoza, Tunuyán, Atuel, and Diamante rivers, allowing for the agricultural cultivation of the surrounding Mendoza region. Records attest that vines were well established in Mendoza by the sixteenth century, and while wines were made for local consumption, word spread as far as Buenos Aires that Mendocino wines were very good. Eighteenth-century records attest to a commercial wine trade with Buenos Aires.

Jancis Robinson, in her celebrated *Oxford Companion to Wine*, observes that the traditional explanation for rapid expansion of viticulture was that the Catholic conquerors required a ready supply of sacramental wine, and that missionaries therefore played a central role in establishing vineyards. She notes, however, that there is little evidence to support this view and that most of the early vineyards and winemaking attempts were secular. She also points out that the economic dynamics, including the cost of importing wine and the difficulties of transporting it over land, resulted in the early Spanish conquerors' realizing that they would need to establish local vineyards if they were to continue to enjoy their wine.

All these efforts came to the notice of the imperial powers in Spain. To protect domestic wine producers and merchants, the Spanish restricted wine production in their New World possessions, effectively freezing any further secular development of viticulture in Mexico. This move inadvertently provided an incentive to Peruvian producers, who soon became the dominant wine suppliers to most of South America. By the seventeenth century, Jesuit missions along the coastal valleys of Peru had become the most important centers of viticulture in South America and remained so until other regions developed their own industries.

The arrival of grapes in Brazil was slightly more recent. Brazil was claimed by Portugal in April 1500 with the arrival of a Portuguese armada of thirteen ships and more than one thousand men, commanded by Pedro Álvares Cabral. He disembarked at what is known today as Porto Seguro in northeastern Brazil. In 1532, the first vines were planted by Martim Afonso de Sousa, in São Vicente in the southeast, but they did not thrive. He had greater success in 1551 in Tatuapé, São Paulo. When the Jesuits arrived in southern Brazil in the 1620s, Roque González de Santa Cruz, with the labor of the native Guarani community, expanded and maintained the vineyards in Rio Grande do Sul.

The history of winemaking farther south, in Uruguay, is more obscure. Although there are local records documenting that Spanish settlers brought the first grapevines to Uruguay during the colonial period, there are no further references to viticulture in the region until the end of the nineteenth century.

In the north and east, viticulture was introduced to Venezuela in the 1520s by the Spaniards in the cities of Coro (the capital of the state of Falcón and the oldest city in western Venezuela) and Cumaná (the capital of the state of Sucre, east of Caracas). Wine production in Bolivia was introduced by the Spanish and Portuguese in the late fifteenth century, with the first vines being planted in Mizque, in the department of Cochabamba, and later extending to Camargo in Chuquisaca. In 1584, grape production finally arrived in Tarija, currently the largest grape-producing region in Bolivia. I could find no definitive written history for viticulture in either Colombia or Ecuador, but presumably their plantings of vines have similar origins.

MODERN LOVE

The modern evolution and reshaping of wine culture among the continent's "big four" producers (Argentina, Brazil, Chile, and Uruguay) are intriguing stories. As in Spain before the Franco era, the wine producers of Chile, Argentina, and Uruguay focused on their domestic markets. In Chile and Argentina, this was in part a result of social and political factors that contributed to closed markets until the 1990s. But with a sharp decline in the regional consumption of wine and the formation in 1991 of the Mercosur free-trade association and its subsequent strategic wine plan, growers across the continent were forced to expand their view to include international markets and to redefine their quality standards.

Argentina, Chile, and Uruguay had very healthy per capita wine-consumption figures going into the 1980s. The average Argentinean consumed more than 90 liters of wine every year, Chileans were drinking 55 liters apiece, and Uruguay's population enjoyed a healthy 33-plus liters. By the late 1980s those numbers had plummeted: Argentina was down to less than 50 liters per person, Chile to 25, and Uruguay to 30. Contemporary per capita consumption remains much lower than in the 1980s, at around 26 liters in Argentina, 17 liters in Chile, and 26 liters in Uruguay.

The reasons for this downward spiral included increasing consumption of other beverages (such as beer, spirits, and sodas); economic crises; and, in Argentina and Chile, government policies that did not favor wine producers. Draconian reforms of drinking and driving laws effectively restricted market growth. Faced with domestic markets that were no longer able or willing to gulp down all the wine that was being made, producers had to consider alternatives.

In Argentina, which even now exports less than 30 percent of its wines, these adjustments entailed changing wine styles to win over foreign palates while making the domestic market more sophisticated. Producers in Brazil and Uruguay did the same. In Chile, the need for change was more urgent, as 70 percent of the country's wine was produced for overseas consumption and hence needed to be reshaped into export-friendly styles. In all four countries, much of this change was accomplished by varying the grape selections. However, in the end, all required outside help to revamp their wine industries.

Chile's success as a modern wine exporter is widely credited to the arrival of Miguel Torres from Spain's Penedès region. At the wish of his father (Miguel Torres Sr.), he explored Chile, accompanied by the Chilean enologist Alejandro Parot (a classmate at the University of Dijon). He was so impressed with the natural conditions that he discovered—a diversity of microclimates and a relative absence of vine diseases—that he purchased an old winery in Curicó in 1979, calling the area a "viticultural paradise." Recognizing an immediate need to raise the game, he imported stainless steel tanks for temperature control and cold fermentations, oak barrels, and other equipment from Europe and went to work in the vineyards planting and caring for French varieties. He soon made an indelible mark on Chilean viticulture. Encouraged by his success and by government policies that were opening up the market, other European-sponsored enterprises soon followed, among them Los Vascos (a joint venture established by Chile's Eyzaguirre family and the Rothschilds of Bordeaux), Hacienda Araucano (established by Bordeaux's eminent Lurton family in 1988), and Lapostolle (the Chilean vision of the Marnier-Lapostolle and Rabat families). This enormous period of advancement is referred to as "stage 1" by Rafael Guilisasti, vice chair of the board of the huge Concha y Toro wine company and a proprietor of Emiliana Vineyards. Stage 2, from the mid-1990s to the early 2000s, was marked by a planting explosion that doubled the acreage under cultivation (from 134,400 acres to more than 270,000 acres) using better grape varieties and clonal selections. Also crucial to this expansion was the discovery of new growing areas, spearheaded by Pablo Morandé's pioneering entry into Casablanca in 1982. This encouraged pushes by others into San Antonio, Elquí, and the cooler subregions of traditional areas such as Colchagua and Maipo. Subsequent advances in viticulture have been led by vineyard mavens such as the biodynamic guru Álvaro Espinoza and the terroir whisperer Pedro Parra. As a result of these developments, Chile is experiencing what we might call stage 3 in its progress toward a bright wine future.

On the other side of the Andes, the situation in the 1970s and early 1980s was equally bleak. Argentina's military government had incurred large amounts of debt with botched projects, corrupt political dealings, the Falklands War, and the state's expropriation of private entities. By the time democracy was restored in 1983, unemployment was skyrocketing, wages were plummeting, and inflation was on the rise. Argentina simply didn't have the means or leadership to recover. In 1989, with a declining gross domestic product and inflation in excess of 5,000 percent, the economy collapsed. This collapse dramatically slowed wine consumption and led to a substantial government-subsidized pull-up of older vineyards. Sadly, this effort led to the destruction of much of the stock of old-vine Malbec, a loss from which the Argentinean wine industry has taken decades to recover.

The Catena family of Mendoza is correctly credited with helping to revive the industry. Nicolás, the patriarch of the clan, spent time at the University of California, Berkeley, in the late seventies. During numerous visits to the Robert Mondavi Winery, he shared quality time with the late Robert Mondavi and learned the growing and winemaking techniques of the Napa Valley. When he returned home to take over the family winery in the eighties, he was determined to build Mendoza into an internationally renowned wine region. He embarked on this effort with the assistance of the renowned winemaker and consultant Paul Hobbs, a classmate of his brother Jorge at UC Davis. With Paul's insistence and Nicolás's persistence, things changed. They harvested smaller crops, restrained vineyard vigor, reduced and carefully timed vineyard irrigation, and implemented cleaner practices in the winery and the cellar. Hobbs became so enamored with the process that he put his money where his mouth was and started his own winery, Viña Cobos, with two Argentinean partners, Luis Barraud and Andrea Marchiori, in 1998. He continues to consult for a dozen or so Argentinean wineries and imports several lines into the United States.

While Nicolás Catena was getting his bearings in Mendoza, Arnaldo Etchart of the Salta region was determined to improve his red wines by tapping the skills of the Bordelais wine expert Michel Rolland. The impact of Rolland's 1987 visit went beyond anyone's expectations. Like Paul Hobbs, Rolland was inspired to invest heavily in Argentina. His projects range from San Pedro de Yacochuya, a Salteña project he runs with Arnaldo Etchart, to his game-changing Clos de los Siete project, involving several wineries in Mendoza. Together, Hobbs and Rolland, along with the Catenas and Etcharts, have led the regime change in Argentinean winemaking.

In Brazil and in Uruguay, the governments were among the principal players responsible for moving the industry beyond *viño común*, the "common wines" or table wines of the past, to the fine wines that are increasingly becoming renowned and opening the markets to export. Unlike Chile and Argentina, where vineyards are planted with a lot of traditional European wine grapes (species *Vitis vinifera*), Brazil and Uruguay are still planted predominantly with North American native (*Vitis labrusca*) and hybrid grapes. Introduced at the end of the nineteenth century, these varieties still account for over 70 percent of Brazil's production and a similar percentage of Uruguay's. In an effort to ameliorate quality and open the markets, both countries went through government-endorsed and partly government-subsidized "reconversions," in which vineyards were dug up and replanted with different vines.

In Brazil, this reconversion was nominally voluntary, although in practice it was essentially forced upon the industry. Until the late 1960s, most of Brazil's wine was common wine, produced by cooperatives. They used the American native and hybrid varieties, which were commercially popular for their ability to produce abundantly in Brazil's climate and their multiple uses (juices, concentrate, jellies, etc.), and the only winemaking expertise was domestic. In the ensuing years, wine companies from France (Pernod, Moët & Chandon) and Italy (Martini and Rossi, Cinzano) entered the market and replanted vineyards with quality *vinifera* grapes. In the years that followed, the combination of a tough economy and political policies began to force the hand of the grape growers, making selling grapes to the co-ops less appealing. Some growers started to work with Brazilian wineries that were expanding and focusing on quality (such as Salton, Miolo, and Casa Valduga), while others decided to try producing their own wines. It was these independent growers and producers who took the first steps toward establishing a premium wine industry, even sending their children abroad to take enology courses. Led by visionaries such as Adriano Miolo, Lidio Carraro, and Flavio Pizzato, Brazil's industry began to reinvent itself in the late 1990s and early 2000s and showed the potential to produce world-class wines. Though fine wines represent only a small portion of Brazil's output, they are welcome entries on the world's wine shelves and are becoming more widely available.

In Uruguay, although domestic wine consumption expanded from the 1950s until the late 1970s, it had dropped precipitously by the 1980s. Faced with a domestic decline in consumption as well as the prospect of the more open international market of the Mercosur, producers saw that they had two options: quit, and cede the market to competitors from Chile and Argentina, or else boost their quality and try to win back the domestic market. As in Brazil, this decision resulted in "reconversion." In the late 1980s and early 1990s, almost all growers and wineries took advantage of government incentives to pull up inferior *Vitis labrusca* grapes and replant with *Vitis vinifera*. However, the benefits of these programs were limited because the government sponsored improvements for fruit only—not for rootstock replacements or any winery improvements. Nevertheless, the policy had a momentous effect on Uruguayan wine production. Led by the energetic and determined Reinaldo de Lucca, who went to Montpellier to study enology (and who today owns a bodega bearing his name), better grape clones, starting with Tannat (the signature grape of Uruguay), were brought over from France, quarantined, heat-treated against viruses, and planted. De Lucca still runs a nursery that promotes Uruguayan grape diversity, and he is the country's leading importer of plant material.

Uruguayan winemaking entered a second phase of improvements as an increasing number of producers committed themselves to quality production, including Daniel Pisano, Francisco Carrau, and Fernando Deicas. Although there are fewer than two dozen wineries today that concentrate on exports, and perhaps twice that number focusing exclusively on *vinifera*-driven quality wines, the momentum is palpable.

Modern refinement of grape growing and winemaking in South America has followed a track parallel to that of other regions that emerged from introversion and obscurity: Spain, South Africa, Portugal, and even Australia in the era before Rosemount and Penfolds. The chapters that follow explore these developments in more depth.

2

GRAPE VARIETIES

Man harvesting grapes in a traditional straw hat (*chupalla*) at Luis Felipe Edwards winery in Colchagua, Chile. Photo by Matt Wilson.

Though we tend to assume that grape growing in South America is limited to a handful of varieties, the continent has a vast selection of grapes that reflects the diversity of its settlers and immigration patterns over several centuries. Official sources indicate commercial plantings of 165 different grapes in Argentina, 117 in Brazil, 65 in Uruguay, and over 60 in Chile.

The origins of these grapes can be traced to Europe: specific cultivars emanate from Spain (including the criolla family of grapes that were first brought over by the conquistadores and early explorers in the late fifteenth and sixteenth centuries), Italy (especially varieties found in Argentina and Brazil), France (whose strong influence can be felt in Chile, which in turn shared it with Argentina, Brazil, and Uruguay), and Portugal (mostly in Brazil). South America is also home to abundant quantities of native North American and North American hybrid grapes. These form the backbone of the Brazilian and Uruguayan table-wine industries and are also used for grape juices, jellies, jams, table grapes, raisins, and concentrates.

In addition to the producers mentioned below, further recommendations can be found in the individual country chapters and the "top ten" lists in the back of the book.

WHITE VARIETIES

CEREZA. *See p. 35*

CHARDONNAY

> *Grown in:* Chile (26,770 acres), Argentina (almost 16,000 acres), Brazil (1,600 acres), and Uruguay (355 acres); small amounts are planted in Peru, Bolivia, Venezuela, Ecuador, and Paraguay
>
> *Distinctive in:* Argentina (Mendoza: Uco Valley), Chile (Aconcagua: Casablanca Valley, San Antonio Valley; Coquimbo: Elquí Valley, Limarí Valley; Southern Regions: Bío-Bío Valley, Malleco Valley), Brazil (Campanha Gaúcha; Serra Gaúcha: Pinto Bandeira, Vale dos Vinhedos; Serra do Sudeste: Encruzilhada do Sul), Uruguay (Canelones)

Originating from France's Burgundy and Champagne regions, this grape is found throughout South America and, as elsewhere in the world, is capable of producing a range of styles, both still and sparkling, that range from the pedestrian to the sublime. Finding the right terroir while judiciously managing the use of oak (used for fermentation and aging) is key for this "queen of whites." Few grapes have witnessed as much improvement in South America as Chardonnay, as new, cooler-climate areas have been discovered and planted.

Until a few years ago, Chile was infamous for its boatloads of correct, mainstream, but uninspiring Chardonnay—striking the two notes of ripe fruit and ample oak. However, as cooler spots have been identified, quality is improving. Today's winemakers are keen on Limarí, principally in the unofficial region of Quebrada Seca around 625 feet above sea level, where clay-rich limestone soils give elegance, verve, and a kiss of salinity. Wines from Maycas del Limarí and Tabali are representative. Equally noteworthy areas include San Antonio (whose wines are waxier and riper in style) and the cooler coastal subregions of Casablanca, where citrus flavors and zesty acidity make for a great framework: good examples come from Viña Casablanca, Loma Larga, and Emiliana. Farther south, the cooler continental spots, including Bío-Bío and Malleco, are exciting. Look for bottlings from Llai Llai (Bío-Bío) and Viña Aquitania (Malleco).

There's ample Chardonnay over in Argentina as well, and some of it is quite good, especially in the Uco Valley, with its outcroppings of limestone. Complexity can be found in locations like Gualtallary and Altamira, exemplified in Catena Zapata's "White Stones." Closer to Mendoza proper, Chardonnay is less consistent, though correct. It's worth keeping an eye on Salta, as the high altitudes, mica soils, and wide variations in day and night temperatures (known as *thermal amplitude*) could make for some excellent wines. Brazil's cool-climate Santa Catarina has recently demonstrated the ability to make crunchy Chardonnays with mineral notes and minimal oak. Encruzilhada do Sul, a few hours south of the Vale dos Vinhedos, is home to equally zippy wines with more green-apple and lemon-lime notes. Pinto Bandeira is noteworthy for the quality of its Chardonnay, much of which is used to make the first-rate sparkling wines of the greater Serra Gaúcha. Finally, in Campanha, Brazil, Chardonnays tend to be fleshier and riper. Lidio Carraro and Don Giovanni are representative Brazilian producers. In Uruguay, a few good efforts are being made in the greater Canelones region, as demonstrated by wines from Marichal and H. Stagnari.

CRIOLLA GRANDE. *See p. 35*

GEWÜRZTRAMINER

> *Grown in:* Chile (790 acres), Brazil (120 acres), Uruguay (60 acres), Argentina (50 acres)
>
> *Distinctive in:* Chile (Aconcagua: San Antonio), Brazil (Campanha Gaúcha)

As stated by Jancis Robinson and her team in the magnum opus *Wine Grapes*, this grape has been proved to be a variant of the Savagnin Blanc grape, from Tramin in northeast Italy's Alto Adige. It likely arrived in South America with the Italians in the late 1870s, and more recently vines were brought directly from France. South American Gewürztraminer, often called or labeled Traminer, is generally milder and less aromatic than offerings from Alsace and North America.

Perhaps one of the last grapes one would associate with South America, Gewürztraminer is capable of making wines with the distinct scents of lychee, rose petal, and honeysuckle that many people love. Though lacking the intensity of classic examples from Alsace, they can nevertheless be perfumed and quite interesting. Look for bottles from Chile's Casa Marín and Matetic. Examples from Brazil's Cordilheira de Santana, in Campanha Gaúcha's Paloma region, are standouts that come closest to the Alsace style and are both delicious and age-worthy.

MUSCAT

> *Grown in:* Argentina (10,100 acres, mostly Alexandria and Rosado), Chile (8,125 acres, mostly Alexandria), Brazil (2,835 acres), Peru (892 acres), Uruguay (35 acres), Bolivia (no acreage data available)
>
> *Distinctive in:* Chile (Coquimbo: Elquí; Atacama: Copaipo, Huasco; Southern Regions: Itata), Brazil (Serra Gaúcha: Farroupilha, Alto Feliz, Vale dos Vinhedos)

This globally prolific grape flourishes in South America, where the range of strains and variants is enormous. The primary examples found include Muscat of Alexandria, used mostly for the production of Chilean and Peruvian pisco and for *singani* (Bolivia's equivalent spirit), and Muscat à Petits Grains, the superior version of white Muscat found mostly in Brazil and used for sparkling wines. In addition there are abundant quantities of black Muscats, including Muscat of Hamburg (Black Muscat) and Muscat Bailey A (Muscat Hamburg crossed with Bailey, a native American grape), which originated in Japan. Rosado Moscato is found in Argentina and Brazil (likely originating from Italy's Trentino). Other white Muscats include Moscato Giallo and Moscato R2, a highly aromatic Italian selection, both of which are grown in Brazil. Finally, a small amount of Muscat Ottonel (a Chasselas-Muscat cross) is found in Uruguay.

Almost all Chilean plantings, vinification, and distillation of Muscat fruit are located in the north, in the Coquimbo region. In the neighboring region of Atacama, Muscat is grown in Huasco and Copaipo purely for pisco. Muscat is also found in the Southern Regions, especially Itata.

Brazil is where Muscat really shines, mostly in the form of sparkling wines (*espumante* and the less effervescent *frizzante*). The finest white Muscat is grown in Farroupilha, which accounts for more than 50 percent of the country's Muscat plantings. At an altitude of 2,600 feet, this region has high thermal amplitude and a long growing season. In Brazil, there are three prominent types of white Muscat: Moscato Bianco, the most widely planted; Moscato Giallo, used for both still and sparkling wines; and Moscato R2. Black Muscats, most notably Muscat Hamburg, are the source of both rosé sparkling wines and a few dessert wines that are occasionally fortified with brandy to make *vins doux naturels*. Noteworthy Brazilian Muscats can be found at Perini, Aurora, and Courmayeur.

Argentina's Muscats are used for the production of basic table wine, juice, and some table fruit. Several Muscat Ottonels from Uruguay are quite enjoyable, especially Ariano's.

NIAGARA

> *Grown in:* Brazil (7,015 acres of white and 1,800 acres of rosé in Rio Grande do Sul; São Paulo [no acreage data available]), Uruguay (approximately 300 acres combined)
> *Distinctive in:* Brazil (São Paulo, Rio Grande do Sul)

One of the few white North American grapes found in South America, this low-acid, foxy (read "funky") grape is found in both white and rosé variants. Based on a cross between Concord and Cassady in the late 1800s in New York State, it made its way to South America in 1894 with merchants who settled in Brazil.

Niagara has become a popular variety in Brazil, especially in the state of São Paulo, where it is the most widely cultivated variety; the Pink Niagara variant makes up 60 percent of the region's total grape acreage. It is also grown in quantity in the state of Rio

Grande do Sul, where it's the third most widely planted variety. Niagara is grown for common wine, juice, and table fruit.

PEDRO GIMÉNEZ. *See p. 36*

PEDRO GIMÉNEZ. *See p. 36*

PINOT GRIS/PINOT GRIGIO

Grown in: Argentina (730 acres), Chile (635 acres), Uruguay (30 acres), Brazil (7 acres)

Distinctive in: Argentina (Mendoza: Uco Valley)

A mutation of Pinot Noir originating in France's Burgundy (where it's rarely found today), this pleasant and tasty grape is the source of enjoyable wines in Alsace, Italy, Germany, and the New World. Currently not much of this variety is planted in South America, but there is increasing interest in the grape and its potential.

As the world has developed a thirst for Pinot Gris/Pinot Grigio, so too has South America. It has shown some success, especially in Mendoza's Uco Valley, where François Lurton's efforts are widely applauded. In Argentina, as in other parts of the world, this grape has notes of pear, honeysuckle, light yellow apple, and white flowers. Look for more of this grape in Chile's Limarí and perhaps San Antonio (just a hunch).

PROSECCO/GLERA

Grown in: Brazil (180 acres), Argentina (22 acres)

Distinctive in: Brazil (Serra Gaúcha: Vale dos Vinhedos, Vale Trentino)

The grape that has become synonymous with chic and modestly priced bubbly originating from the eponymous northern Italian wine region is also found in Brazil. Since 2009, the use of the name *Prosecco* has been limited by law to the Veneto; elsewhere the variety is now labeled as Glera, the name used in Trieste. In Brazil, as in Australia, the use of the term *Prosecco* is still allowed on labels as long as the wine is not exported.

I was surprised to discover that Prosecco is as tasty in Brazil as in Italy. Putting forward subtle notes of ripe pear, bitter almond, citrus blossoms, and candied citrus peel, Brazilian Prosecco is not the global juggernaut we find in North American restaurants but still performs well. Almost all the fruit is cultivated in the Serra Gaúcha's primary regions. If it weren't for the large amounts of sparkling Muscat and well-made Brazilian bubblies from Chardonnay and Pinot Noir, Prosecco might be even more popular domestically. When you're there, try bottles from Garibaldi, Perini, and Peterlongo, which pair sublimely with the deep-fried *croquetes* made with chicken or cheese.

Grown in: Chile (820 acres), Argentina (219 acres), Brazil (70 acres, mostly Italico), Uruguay (42 acres)

Distinctive in: Chile (Aconcagua: San Antonio; Southern Regions: Bío-Bío)

Riesling and Riesling Italico are two different varieties. Riesling Italico is also known as Graševina (the leading white variety in Croatia). Riesling, the noble white variety famed in Germany and Austria, has proved it can also excel in the New World, as demonstrated by wines from Australia's Eden and Clare valleys and from Washington State, Michigan, and New York State. In South America, the best examples have thus far originated in cool-climate Chile, although growers are also excited about its potential in cooler areas of Brazil and Uruguay. The Italico grape yields some enjoyable and fresh wines from Brazil, though with less perfume and aroma than classic Riesling.

Currently grown in just a few regions, Chilean Rieslings are elegant, though restrained, and weightier than their Old World counterparts. The best efforts seem to be coming from San Antonio, where the styles range from off-dry to dry and are punctuated with floral, sweet citrus, and some stony notes. In Bío-Bío, they have a lifted perfume and are limey and occasionally petrol scented. Casa Marín and Matetic are exemplary. In Brazil (and in a small way in Uruguay), Riesling Italico is productive and capable of making some intriguing wines that, at their best, have floral and slightly exotic fruit notes and bright acidity. They can be dry or off-dry. Aurora and Lovara make benchmark examples.

SAUVIGNONASSE/SAUVIGNON VERT

Grown in: Chile (2,115 acres), Argentina (1,300 acres, mostly in Mendoza)

Distinctive in: Chile (Central Valley: Curicó, Maule)

A variety with lots of historical baggage, in Chile this grape was long confused with Sauvignon Blanc, giving the country a reputation as an underperformer with the latter variety until the grape's true identity was revealed. Still claiming a lot of acreage, much of it in Curicó, the grape has no direct connection to the nobler Sauvignon Blanc but is related to the Friulano grape of northeastern Italy. Its arrival in South America, however, can be attributed to Bordeaux: planted in the vineyards there along with Sauvignon Blanc and Semillon, it hitchhiked to the New World with these varieties. Also known as Sauvignon Vert in Chile, it is produced in high volumes (some estimates claim that it accounts for 85 percent of all Sauvignon planted in Chile) and is rarely of distinctive quality.

Sauvignon Vert can have a fresh character when young but is light on the palate and takes on a dried, vegetal character not long after bottling. Found mostly in south-central Chile (especially in Curicó), it's known for a very pungent and oniony, musky, or feral character. When used in small amounts—for example, when blended with good-quality Sauvignon Blanc—this can be attractive. At worst, it's like body odor. In Argentina, its role is similar.

SAUVIGNON BLANC

> *Grown in:* Chile (32,800 acres), Argentina (5,670 acres), Uruguay (363 acres), Brazil (190 acres), Bolivia (86 acres)
>
> *Distinctive in:* Chile (Aconcagua: Casablanca, San Antonio; Coquimbo: Limarí; Central Valley: Rapel [Colchagua]), Argentina (Mendoza: Uco Valley, San Rafael), Uruguay (Maldonado)

One of the wine world's most successful expatriates, France's gift from the Loire Valley and Bordeaux is a star performer in many parts of the Old World (including Italy and Spain) as well as in the United States, South Africa, Australia, New Zealand—and now Chile. Arguably Chile's strongest white wine grape, it displays clear regional nuances. Sauvignon Blanc is also capable of producing very good wines in Argentina, Uruguay, and occasionally Brazil. Small amounts are grown in Bolivia.

The first correct and certified Sauvignon Blanc clones are said to have arrived in Chile in the 1980s from the University of California at Davis, and more subsequently came from France. In concert with experimental planting in coastal areas such as Casablanca and San Antonio, quality has improved. Pablo Morandé's 1980 arrival in Casablanca ushered in a new wave of wines, with a cleaner, brighter perfume and the definitive Sauvignon characters of herbs, grass, lemon, and olive. Today, the western, cooler-climate area of Casablanca produces zippier versions with notes of lemongrass, while farther east the wines tend to evoke pink grapefruit, hard candy, and green melon. Great examples of Casablanca Sauvignon Blancs are produced by Cono Sur, Casas del Bosque, and Errázuriz. In San Antonio, the wines have an oilier, yellow-plum character in addition to sorrel, basil, and lemon-thyme notes. Examples from the subregions of Leyda and Lo Abarca are more concentrated and still spicy, but with nettle and softer elderflower notes, as in the bottlings from Casa Marín, Viña Leyda, and Undurraga's "Terroir Hunter" series. Other new areas for Sauvignon Blanc are emerging, including Limarí (where the wines are mineral driven) and Bío-Bío (elegant with quince aromas).

In Argentina, the best efforts are made in Mendoza's Uco Valley, especially in Tupungato, where the wines have aromas of citrus, cut grass, and verbena. Check out those from O. Fournier and Salentein.

Sauvignon Blanc is no more than a bit player in Brazil. In Uruguay, however, it is the leading white grape for fine wines, representing close to 53 percent of all quality white-variety plantings. Excelling in all three coastal regions (Canelones, Colonia, and Maldonado), Sauvignon can be bright citrus and mineral (Maldonado), reminiscent simultaneously of France's Sancerre and New Zealand's Marlborough (Colonia), or with notes of olive and grapefruit or pomelo (Canelones). Standouts in Uruguay include Castillo Viejo, Bodegas Carrau's "Sur Lie," and Juanico's Familia Deicas "Atlántico Sur."

SAUVIGNON GRIS

Grown in: Chile (346 acres), Uruguay (20 acres), Argentina (14 acres)

Distinctive in: Chile (Aconcagua: San Antonio)

This rare variant of Sauvignon Blanc is hard to find, but when you hit an example you are inevitably happy that you did. Interestingly, the wines are pinkish-beige rather than gray, as the name would suggest.

Sauvignon Gris is prized for lifted aromatics and a rich and round smoothness. Although historically cherished, it has fallen out of favor because of its very low yields. When well made, the wine is full of ripe white stone fruit, pink grapefruit, yellow apple, and pear. The wines have a creamy, almost velvet mouthfeel with a soft, persistent, earthy grit. In Chile, some tasty bottlings come from San Antonio, especially Leyda, and Maipo. Examples from Uruguay are especially exciting, despite the small quantities produced, such as Filgueira's and Bodegas Carrau's "1752."

SEMILLON

Grown in: Chile (2,295 acres), Argentina (2,110 acres), Brazil (400 acres), Uruguay (59 acres)

Distinctive in: Rarely distinctive anywhere

Like many other grapes, this variety came over to Chile by way of Bordeaux in the nineteenth century and then crossed the Andes to Argentina and beyond. Less prominent in South America today, it is used mostly as it is in France, as a blending grape with Sauvignon Blanc and with other whites. It rarely displays enough complexity to be worth bottling as a varietal wine.

A few winemakers appreciate the grape for its historical significance and still make delightful wine from it (Roberto de la Mota at Argentina's Mendel being a case in point). Otherwise, the best examples tend to be dessert wines.

TORRONTÉS. *See p. 37*

VIOGNIER

Grown in: Argentina (2,016 acres), Chile (1,940 acres), Uruguay (112 acres), Brazil (17 acres)

Distinctive in: Argentina (Mendoza: Uco Valley [Tupungato], Luján de Cuyo), Chile (Central Valley: Rapel [Cachapoal, Colchagua]; Coquimbo: Limarí), Uruguay (Canelones)

This highly perfumed and aromatic grape has created a lot of buzz in South America. Long neglected while winemakers focused on Chardonnay and Sauvignon Blanc, Viognier has been evoking new interest, and there seems to be more of it served up each year. The grape generally exhibits a far more muted nose in South America than in other New World locations (such as the United States and Australia), but it's spot-on in the mouth. South American Viognier can be delicious both on its own and blended with other white grapes, such as Sauvignon Blanc.

Argentina's Viogniers break out into two styles, which seem to be associated with different altitudes. Those coming from the Uco Valley have melon, apricot, guava, and kiwifruit notes in a slightly leaner style, while those of Luján de Cuyo tend to be waxier, with similar but riper fruit qualities. Chile's versions, especially in Cachapoal, have more classic apricot and peach characters, often coupled with notes of orange blossom and honeysuckle in riper vintages. Though not life changing, Argentinean and Chilean Viognier-based wines are almost always better value for money than other global examples. Viognier's ability to fix color when cofermented with Syrah, as is done in the Côte-Rôtie in France, is increasingly being applied in Chile and in Argentina (where the same is also done with Malbec, with some intriguing results). In Uruguay, the grape is very vigorous, so early removal of about half the unripe fruit is important to give the wines any varietal character. Some of the nicer examples include Argentina's Doña Paula, Finca Sophenia, and DiamAndes and Chile's Casa Marín, Cono Sur, and Concha y Toro's "Lo Ovalle." In Brazil, look out for Kranz and RAR; in Uruguay, try Irurtia and Bodega Garzón.

RED VARIETIES

ALICANTE BOUSCHET

Grown in: Chile (764 acres), Argentina (531 acres), Brazil (225 acres), Uruguay (54 acres)

Distinctive in: Brazil (Serra Gaúcha: Vale dos Vinhedos)

Although this inky-black variety originated in France as a cross between Petit Bouschet and Grenache, it seems more at home speaking Portuguese: it excels in Portugal's Alentejo region and in Brazil's Serra Gaúcha, where it is championed by the winemaker Flavio Pizzato. It is one of the *teinturier* grapes, red grapes whose flesh and juice are red (due to anthocyanin pigments) rather than clear, like the flesh of other red varieties.

You hear about Alicante frequently, but not as a varietal wine: typically the grape is used in blended red wines, where it is valued for its color, tannin structure, and ability to add complexity to simple, fruity reds. It deserves a chance in the limelight, as the inherent spicy and peppery, black-fruit compote, cola-nut, and cacao flavors are distinctive, and a red wine containing mostly Alicante can be very good. Besides Pizzato, Dal Pizzol makes a lovely example.

ANCELLOTTA

Grown in: Argentina (745 acres), Brazil (450 acres)

Distinctive in: Brazil (Serra Gaúcha)

Once rarely seen outside Italy, this grape cuts a swath across lower South America. It can be found bottled as a varietal wine in Brazil (which makes sense given Brazil's viticultural connections to northern Italy). In Argentina, it's apparently used solely for blending: I have not encountered it as a labeled varietal wine. As in its native home of Emilia-Romagna (where it is used in lambrusco blends), the grape is prized for its color and structure.

In their book *Wine Grapes,* Jancis Robinson and her coauthors observe that the Brazilians take greater interest in this Italian grape than the Italians do, and I agree. Brazil's best versions taste somewhat like Amarone, a rich, dry red wine made from partially dried grapes from the Valpolicella region of Italy, with concentrated, almost desiccated flavors of mocha, raisins, cherry, and anise seed (though in Brazil the grapes are typically not dried). A few producers, such as Pizzato and Don Guerino, bottle very good varietal examples.

ARINAROA

Grown in: Uruguay (69 acres), Brazil (7 acres), Argentina (2.5 acres)

Distinctive in: Uruguay (Canelones), Brazil (Serra Gaúcha: Vale dos Vinhedos)

Talk about a tongue twister! Try saying it after you have had a glass or two. This cross of Tannat and Cabernet Sauvignon (not Merlot and Petit Verdot, as claimed in the mid-1950s) is not very common in South America, though it adds to the wine industry's diversity there. Its name, as Jancis Robinson points out in *Wine Grapes,* comes from the Basque *arin,* meaning light, and *arno,* meaning wine.

Like Alicante, Arinaroa is rarely bottled as a varietal wine. The best example I've had was at Casa Valduga in Brazil's Serra Gaúcha, where it exhibited mulberry, cassis, licorice, and laurel notes in a tight and lively package. Similarly, in Uruguay, few varietal wines are made, but Giménez Méndez bottles a good example of this grape from a vineyard in Las Brujas, a subdistrict of Canelones.

BARBERA

Grown in: Argentina (1,345 acres), Brazil (75 acres)

Distinctive in: Rarely distinctive anywhere

The stalwart red grape that ranks second in its native Italy for red-grape plantings is also found in South American countries with a connection to the motherland, Argentina and Brazil. However, here it is treated chiefly as a blending variety and is mostly lost in everyday wines.

This is a classic story of quantity trumping quality. One exception from Argentina is the bottling from Bodega Norton, which speaks to all of the grape's lively red-cherry fruit, vibrant acidity, and herbal notes. Some organically grown grapes in La Rioja are bottled under the Picas Negras label of the La Riojana Cooperative, and Altos de San Isidro in Salta produces a 100 percent varietal wine. In Brazil, Barbera is used exclusively for blending.

BONARDA

Grown in: Argentina (44,800 acres)

Distinctive in: Argentina (Mendoza: Eastern Oasis; San Juan: Tulum)

What the Argentineans love and call Bonarda, their second most widely planted quality red grape after Malbec, is not the Bonarda of the Italian Piedmont: it's a different variety. Argentinean Bonarda is what the French call Douce Noir (in its original home of Savoie) and what the rest of the world calls Corbeau—except in the United States, where we call it Charbono. Nicolás Catena's daughter Laura claims that the reason Argentineans decided to stick with the Italian name Bonarda rather than switch to Corbeau is that they didn't want another French-named grape. But while the name and history may be confusing, the wine is certainly straightforward: a high-volume red of ample color and structure even when exceedingly cropped.

Today Bonarda is Argentina's third most widely planted grape, but it was actually number two less than twenty years ago. When wine consumption fell in the 1980s because of political and economic difficulties, almost two-thirds of Argentina's Malbec vines were pulled out. However, Bonarda was retained, for two reasons: first, it was popular with growers because it provided huge yields while retaining its color and remaining very fruit-forward and tasty; and second, it could ameliorate the low-grade criolla varieties like Cereza and Criolla Grande. Although Bonarda is still responsible for ample volumes of everyday drinking wine, winegrowers have learned that when you decrease the leaf canopy, cut production, and control irrigation, this thin-skinned grape can make very good wine. Bottlings from Nieto Senetiner, Altos las Hormigas, and Lamadrid demonstrate its potential. Straightforward and approachable, the grape oozes with black plum and cherry, blueberry, blue flowers, and potpourri. As the winemaker Susana Balbo articulately explains, it "does well where Malbec doesn't because Malbec does not like very warm climates, whereas Bonarda does." Though often sold as a varietal wine, Bonarda blends beautifully with Malbec, Tempranillo, and even Cabernet.

CABERNET FRANC

> *Grown in:* Chile (3,320 acres), Argentina (1,540 acres), Brazil (1,025 acres), Uruguay (825 acres)
>
> *Distinctive in:* Chile (Aconcagua: Casablanca; Central Valley: Maule), Argentina (Mendoza: Luján de Cuyo, Uco Valley, La Pampa), Brazil (Serra Gaúcha: Vale dos Vinhedos, Campanha Gaúcha),Uruguay (Canelones, Maldonado, San José, Paysandú)

It's intriguing that Cabernet Franc, the anchor variety of the right bank of Bordeaux and the Loire Valley, can be so expressive of the very different terroirs of South America. Not much has been planted yet, but you can bet there's more to come.

This is the one variety that performs well in all the wine-producing countries of South America, expressing with flair the classic varietal markers of raspberry, black- and redcurrant, bay leaf, licorice, and marjoram. Chile leads in production of this grape, which is natural given its links to Bordeaux in the nineteenth century. Until recently Cabernet Franc has mostly been a component of *cortes* (blends), but producers have recently begun producing exciting varietal wines. Garage Wine Company, Gillmore, and Von Siebenthal make lovely examples.

Argentina has had similar results, with attractive efforts being vinified in both Luján de Cuyo and the Uco Valley. Some of the better wines are made by Lagarde, Bressia, and Riglos. In Brazil, Antonio Dal Pizzol, who makes an excellent flagship version, told me that Cabernet Franc was originally one of the key *vinifera* grapes introduced to Brazil, but due to poor clonal selection, most died and have since been replanted. New efforts are very good, including Casa Valduga's "Raizes" and Aurora's "Pequenas Partilhas." The grape has shone across Uruguay, too, with many of the best non-Tannat wines being made with this variety, from Maldonado (Alto de la Ballena) to Paysandú (Leonardo Falcone).

CABERNET SAUVIGNON

> *Grown in:* Chile (94,950 acres), Argentina (40,450 acres), Brazil (4,650 acres), Uruguay (1,685 acres), Bolivia, Peru, Colombia, Ecuador, Venezuela

Distinctive in: Chile (Central Valley: Maipo, Rapel [Colchagua]; Argentina (Mendoza: Luján de Cuyo, Uco Valley), Brazil (Serra Gaúcha: Campanha Gaúcha)

The worldwide "king of reds," Cabernet Sauvignon, is a very important grape on this continent. South American winemakers took to heart the early advice from French wine experts that to be taken seriously required a focus on red Bordeaux varieties, and ample effort has been made to establish this prestigious grape. Though there are many excellent examples, even the most supportive fans will tell you that there's still work to be done.

Cabernet Sauvignon is Chile's principal red grape, beating out Carmenère by a ratio of about six to one. Chile's Cabernet vineyards enjoy ample sun, well-drained soils, few problems with pests or ripening, and cool Pacific breezes—a grape nirvana. Too often, however, the ease of cultivation leads growers and winemakers to ask too much from the vines: excessive yields result in diluted flavors, and planting in borderline areas where it doesn't always get warm enough can result in green fruit.

Alto Maipo is the prime region for Cabernet: its wines resemble those of Pauillac in Haut-Médoc, not too fat and redolent with complex notes of black fruit, herbs, and graphite. Historically, the best sites for Cabernet have been in the east, toward the Andes (Puente Alto, Pirque, Totihue). Well-maintained vineyards farther south (Curicó, Maule, Itata) and in more coastal areas (Limarí, Maipo, Colchagua) are starting to show elegant results, in the opinion of the Chilean expert Peter Richards MW. Carmenère and Syrah are excellent blending buddies with Cabernet in Chile, and many of the best wines are *cortes.* Celebrated examples include "Almaviva," "Viñedo Chadwick," Concha y Toro's "Don Melchor," and a number of Cabernet blends by Montes, including "Alpha M."

Argentina's Cabernet is cultivated widely, with the best coming from Luján de Cuyo in and around the Agrelo, Barrancas, and Las Compuertas areas, where full ripening is not the challenge that it can be elsewhere. Great bottles from Luján include those from Viña Cobos's Bramare Marchiori Vineyard, Mendel, Durigutti, Pascual Toso, and several offerings from Catena Zapata. In warmer, riper vintages, Cabernet Sauvignon from the Uco Valley can be fantastic, especially in Altamira, Gualtallary, and Vista Flores. Look for O. Fournier, François and Jacques Lurton's Piedra Negra, and Riglos.

In Uruguay, Cabernet Sauvignon is also difficult to ripen, but one year in four can be magic, according to the Canelones winemaker Carlos Pizzorno. Brazil's challenges with the grape echo those of coastal Uruguay: more consistent efforts are coming from the drier Campanha Gaúcha in the south, on Uruguay's northern frontier, a spot where Cabernet does ripen with some frequency. In Brazil, Boscato is the most reliable region for Cabernet, and in Uruguay De Lucca and Pizzorno rank high.

CARIGNAN

Grown in: Chile (1,275 acres), Argentina (71 acres)
Distinctive in: Chile (Central Valley: Maule)

Carignan, the pride of Chile's Maule region, is witnessing a fairytale transformation. As in other parts of the world, most notably in the Languedoc in France and in Aragón, Montsant, and Rioja in its native Spain, Carignan is being rethought.

The epicenter for Carignan is Chile's Maule (specifically Cauquenes), about 1,400 dry-farmed acres in an area devastated by the 2010 earthquake. The lion's share of it was planted in 1939, after the first serious earthquake and with government assistance. Carignan has never been celebrated anywhere for its charm or stylishness, and so the Carignan vines aged like sleeping giants for decades, valued only for their ability to add depth and value to the local Pais-based blends. In the 1990s, winemakers began realizing the potential of old-vine Carignan, and a number of them formed the Vigno association to improve it. At its best, old-vine Carignan displays the concentrated dusty red and black plum, dry cherry, and baking spices that epitomize the grape. Some of the most classic examples are from Gillmore, De Martino, Garage Wine Company, and Morandé.

CARMENÈRE

> *Grown in:* Chile (23,480 acres), Argentina (140 acres), Brazil (84 acres)
>
> *Distinctive in:* Chile (Central Valley: Rapel [Colchagua, Cachapoal], Maipo [Isla de Maipo], Maule [Cauquenes]; Aconcagua: Limarí)

Considered the signature grape of Chile, Carmenère is today associated more closely with South America than it is with Bordeaux, its original home. The vines were brought to Chile in the mid-1800s, before the European phylloxera outbreak, and planted directly into the ground. Chile's producers make some superb wines with this Bordelais ugly duckling. The right climate and terroir are essential to bring it out of its otherwise herbaceous and stalky-green shell. Limited amounts are planted in Argentina and Brazil.

Carmenère was not identified as a unique variety in Chile until 1994, by the French grape sleuth Jean Michel Boursiquot, but Peter Richards believes that even before then, most Chilean vintners recognized the difference between Merlot (known as Merlot Merlot) and Carmenère (Merlot Chileno). First bottled as a varietal wine in 1994 by Álvaro Espinoza at the Carmen winery, Carmenère is a late-ripening, vigorous grape that thrives in moderately fertile deep soils in an environment that is sunny, warm, and dry. At its best, it can be stunning. In Maipo, a region known for full ripening, it can evoke black figs and tar with complex savory notes, while in cooler Limarí it leans more toward roasted pasilla peppers and black tea leaf. In Colchagua, generally acknowledged as the best region for Carmenère, the wine is all about bitter chocolate, smoky paprika, and black plum. Colchagua's wines are recognized for their backbone and structure (especially in Apalta and Marchigüe), as are those from Maule's Cauquenes to the southeast. Low in acid (especially when harvested late), Carmenère is often acid adjusted or blended with more structured grapes like Cabernet Sauvignon. Some of Chile's finest blended reds incorporate Carmenère, including Almaviva, Neyen, and Álvaro Espinosa's Antiyal. The list of pure Carmenères is long, but some I have noted as consistently good are Montes, Casa Silva, Pérez Cruz, Terra Noble, De Martino, Concha y Toro, and Viña Morandé.

CONCORD AND OTHER NORTH AMERICAN GRAPES

Grown in: Brazil (Isabella [over 27,000 acres], Bordo [15,000 acres], Concord
   [7,300 acres]), Argentina (115 acres of Isabella), Uruguay
Distinctive in: Rarely distinctive

Success with native North American and hybrid grapes—not only Concord but also Isabella, Bordo, Cynthiana, Delaware, and Norton—is South America's biggest vineyard surprise. Nobody else gives much thought to these varieties outside the United States and Canada. However, ample amounts are cultivated, notably in Brazil, where they make up almost three-quarters of the grapes planted, and in Uruguay, where the proportion is close to two-thirds of all plantings. In those two countries these grapes form the backbone of the common-wine industry and are also important for the large volumes of juices, jellies, table grapes, and concentrate produced. While North Americans prefer their grape juices darker purple (witness Welch's), South Americans favor white grape juice and regularly process red fruit to make white juice.

In Brazil and Uruguay, red hybrid grapes are the backbone of the grape industry and the wine industry, but not of fine-wine production. Although in Uruguay a 2008 law banned new plantings of anything but *vinifera* grapes, Isabella and her friends are still dominant. In Brazil's primary region of Rio Grande do Sul, as recently as the 2011 harvest, production of table wines from hybrids is five times that of fine wines, with red table wines making up close to 80 percent of the total. In the state of Santa Catarina just to the north, the ratio is even more extreme, though the total volumes are smaller. While the juices are delightful, I can't say as much about the wines.

EGIODOLA

Grown in: Brazil (130 acres), Uruguay (8 acres)
Distinctive in: Brazil (Serra Gaúcha: Vale dos Vinhedos)

Originating in France in 1954, this is an obscure cross between the grapes Fer Servadou and Negramoll (also known as Mollar). Although Egiodola has rarely achieved greatness in France, it somehow made it as far as South America, and it makes some distinctive and occasionally tannic wines in Brazil. Small amounts of this usually rich but softer style of wine can also be found in Uruguay. The name is Basque for "true blood."

Brazil seems to be a hub for obscure grape varieties. Brazilian Egiodola has the typical semiopaque color and deep raspberry and roasted-chestnut flavors that characterize it in southwestern France, but it doesn't always have the same bite or ferocious tannins. Vinified as a varietal wine by a handful of winemakers, Egiodola is a cult wine to those who love it and an oddity to those who don't. Pizzato's example is a benchmark. The person responsible for developing this grape, Pierre Marcel Durquéty, is also the architect of the Arinaroa grape.

MALBEC

> *Grown in:* Argentina (76,720 acres), Chile (2,540 acres), Uruguay (100 acres), Brazil (60 acres), Bolivia, Peru
>
> *Distinctive in:* Argentina (Mendoza: Luján de Cuyo, Uco Valley; Salta; Patagonia), Chile (Central Valley: Rapel [Colchagua])

If the color purple had a flavor, it would be Malbec. This variety seemed to come out of nowhere to capture the palates of contemporary red wine drinkers. Ironically, the grape that yields the mouth-filling, smooth, and deeply colored red associated today with Argentina is also responsible for wines that are the polar opposite on the other side of the world. In its native southwest France, where the name *Malbec* translates loosely to "bad mouth," the wines are hard, tannic, austere, and usually unripe. Consumers are often shocked when they find out that oft-austere French Cahors is made from the same grape as sublimely complex wines from Argentina. Argentinean Malbecs run the gamut from wines with distinctive regional character to tasty everyday examples.

It's said that Malbec and Argentina are made for each other. Some claim that the Argentinean version, which is softer, riper, and fleshier than its French counterparts, is genetically different, being rooted in vines planted before the European phylloxera epidemic. The most significant influence on the grape is elevation. Higher altitudes (providing more thermal amplitude and ultraviolet exposure, which encourage the development of thicker, more flavorful skins), coupled with dry and sunny surroundings, create perfect growing conditions. Tiered elevations result in different personalities of fruit, with the lower plateaus of Luján de Cuyo and even Eastern Mendoza yielding smoother and richer styles, while the higher reaches (Uco Valley, Salta) provide more structure and a lean elegance. The Uco Valley, with limestone soils and a cooler climate, yields wines of higher acidity that are linear with sour-cherry fruit and deep color. In Luján, the wines have ample intensity and black fruit, weight, and texture, with smoother, creamier tannins. The wines of adjacent Maipú are slightly less intense, with freshness and very smooth, round tannins. These differences are so pronounced that in a post on Jancis Robinson's Purple Pages blog in September 2010, Richard Hemming suggests classifying Argentina's Malbecs by altitude rather than by subregion: "So rather than having Medrano, Barrancas and Cruz de Piedra (all within Maipú), you might have Mendoza altitude bands A to E, and likewise in Salta, etc. In that sense it would echo the recent Riesling sweetness scale, in appearance on the label, at least: with higher altitudes giving lighter, more aromatic styles, and lower ones giving fuller, more fruit-driven wines. It would certainly be easier for consumers to get a handle on."

Some of Argentina's leading Malbec producers (and there are dozens) include Achaval Ferrer, Alta Vista, Catena Zapata, Domaine San Diego, Doña Paula, Durigutti, Mendel, Noemía, and Zorzal. I mention many more later in the book! Malbec in Chile is at its best in dry-farmed, granitic soils, including the cooler parts of Maipo, San Antonio, Maule, and Colchagua. At its finest, it is still somewhat sharper and harder than its neighbors over the Andes. Because Chilean Malbec is less about purity of fruit and more about thyme, rosemary, cassis, and spice, it lends itself to use in classic Bordeaux blends. A few examples to watch for are Luis Felipe Edwards, Viu Manent, and Polkura. Malbec is quite similar in Brazil and Uruguay.

MARSELAN

Grown in: Argentina (20 acres); Uruguay (225 acres), Brazil (30 acres)

Distinctive in: Brazil (Serra Gaúcha: Vale dos Vinhedos, Vale Trentino)

I fell in love with Marselan, one of my new favorite grapes, in Brazil, where several examples were high on my list of noteworthy bottles. A cross of Cabernet Sauvignon and Grenache (developed in France in 1961), at its best it embodies the best of both. This somewhat obscure grape is also grown in France and apparently in China. Though it is also planted in Uruguay, it's at its best in Brazil.

Named after Marseillan, the small town on France's Mediterranean coast where it was first grown, Marselan has the creamy, fleshy texture of Grenache along with the peacock's-tail complexity of Cabernet Sauvignon, which also imparts a trace of tannin to the mix. It has a little of everything: tasty red-cherry fruit, a somewhat flashy mouthfeel, and soft but discernible tannins. Two producers in Brazil—Perini and Pizzato—make lovely examples. In Uruguay, I expected to find more varietal examples, but it seems to be used more in blends. The iconic wine of Juanicó—Familia Deicas "Prelúdio"—contains Marselan, along with five other grapes, but only as a small percentage of the blend; Juanicó makes a varietal wine as well. Other varietal examples include J. Chiapella and the González Santiago family's Gobelet line, but, alas, I haven't tried either of these.

MERLOT

Grown in: Chile (26,300 acres), Argentina (15,520 acres), Brazil (2,700 acres), Uruguay (2,160 acres), Bolivia (74 acres), Peru, Ecuador, Colombia

Distinctive in: Chile (Central Valley: Rapel [Colchagua], Maipo, Maule), Argentina (Mendoza: Luján de Cuyo, Uco Valley; Patagonia), Brazil (Rio Grande do Sul: Serra Gaúcha, Campanha Gaúcha), Uruguay (Canelones, Maldonado)

The familiarity of Merlot has bred contempt in many parts of the world, but this scorn is not always deserved. Some superb Merlot and Merlot-based blends (hello, Petrus!) grace store shelves. Merlot is not as ubiquitous in South America as it is in the United States, France, or Central Europe, and in the late 1980s and early 1990s it was habitually confused with Chilean Carmenère. Today, however, the continent is blessed with several fine examples, particularly from Brazil.

Peter Richards claims that much of the acreage currently registered as Merlot in Chile (30 to 80 percent) is still Carmenère. Because Chilean Merlot has been profitable, wines labeled as Merlot are often in fact a blend of the two, and typically mostly Carmenère. Nevertheless, there are well-made Merlots from Chile, though many are uninspired. At its best, Chilean Merlot expresses ripe red berry, black plum or berry, soft mint, and spice notes, ranging from clove and nutmeg (cool) to cinnamon and ginger (warm), always with vibrant acidity. Great examples are made by Montes, Lapostolle, Santa Ema, and Errázuriz.

In Argentina, most Merlot ends up in red wine blends, usually with other Cabernet family grapes. Again there are noteworthy examples, though they tend to be at the premium end of the market. Paul Hobbs's great-value El Felino line from Viña Cobos and Bodega Norton's "Barrel Select" are emblematic of what can be done.

Brazil's strongest red wines, year in and year out, are Merlots and Merlot *cortes*. With the most winemaking experience and the oldest Merlot vines, producers in Serra Gaúcha have excelled: this is the only region in Brazil where more Merlot is planted than Cabernet Sauvignon. Stylistically, these wines are Old World in structure (like right-bank Bordeaux) but somewhere in between New World and Old World in flavor, with herbal notes layered over New World berry and plum fruit. Salton's "Desejo" and Miolo's "Terroir" are good examples. Uruguay's efforts are similar to Brazil's, though Uruguay's producers have less history to draw on, and so their Merlots are still works in progress.

MOLLAR. *See p. 36*

PAIS. *See p. 36*

PETIT VERDOT

> *Grown in:* Chile (1,555 acres), Argentina (1,240 acres), Uruguay (50 acres), Peru
> *Distinctive in:* Chile (Central Valley: Rapel [Colchagua]), Argentina (Mendoza: Luján de Cuyo)

An inky and intense grape, Petit Verdot is known around the world as a great grape for adding color and structure to Bordeaux-style red blends. A little bit typically goes a long way, as this grape is simply too powerful and one-dimensional to be bottled solo. However, in South America, it has shown potential for bottling as a varietal wine.

In Chile and Argentina, Petit Verdot is capable of making ripe and not overly astringent varietal wines, with notes of ink and iodine, blackberry extract and violets. Petit Verdot could be a prominent niche player in the next few years; Chile's Von Siebenthal and Argentina's Finca Decero demonstrate its potential.

PINOT NOIR

> *Grown in:* Chile (8,200 acres), Argentina (4,450 acres), Brazil (480 acres), Uruguay (140 acres), Peru
>
> *Distinctive in:* Chile (Aconcagua: Casablanca, San Antonio; Coquimbo: Limarí; Southern Regions: Bío-Bío); Argentina (Patagonia: Neuquén, Río Negro), Brazil (Rio Grande do Sul; Serra Gaúcha: Vale dos Vinhedos, Vale Trentino, Campanha Gaúcha); Uruguay (Colonia: Carmelo)

I suspect that all winemakers dream—publicly or privately—of making a great Pinot Noir before they die. This mission has led to worldwide quests for the optimal climate (not too hot, not too cool) and to terroirs that afford the greatest opportunities for success. In South America, things are no different, although great examples of Pinot Noir are relatively rare. Growers simply do not have the established older vines and diversity of plant material and clonal selections that lead to spectacular wines. And identifying the best spots is very much a work in progress. As in other parts of the world, the grape is also used to make very good bubbly, especially in Brazil.

Winemakers in Chile are still trying to define Pinot Noir. Its lack of distinction there is perhaps driven by a combination of young vines and the fact that most of the Pinot selections are from older Valdivieso clones brought over by Concha y Toro, before the arrival of the superior Dijon clones. Obtaining diversity in plant material in Chile remains a problem and an obstacle to wine quality. Pinot Noir performs well in the Leyda region of San Antonio, displaying more herbs and pomegranate; in Casablanca, the flavors are more red cherry and red raspberry. There is buzz in volcanic Bío-Bío on the hillside area near the Bureo River and in cool Limarí: wines from both areas exhibit earthiness and minerality. Cono Sur, Viña Ventisquero, Veramonte, and Kingston Family are reliable Pinot producers.

In Argentina, the buzz over Pinot Noir is concentrated in Patagonia's regions of Río Negro and Neuquén. Vine age is critical to this variety, and the old vines of the Upper Río Negro Valley subregion produce depth and richness, as bottlings of Humberto Canale and Chacra attest. The promise of the younger Neuquén plantings is illustrated in wines from Familia Schroeder and Fin del Mundo. Mendoza's cooler Uco Valley can also yield successful wines, such as examples from Luca and Mariflor. A lack of diversity in clonal material has been problematic here, too, though less so than in Chile. Brazil's Pinot Noir is mostly used in sparkling wines, either alone or blended with Chardonnay; these can be superlative. Outstanding still wines are rare, but cooler areas, like Santa Catarina and the Serra do Sudeste, show promise, as evidenced by Lidio Carraro's

Dádivas bottling. In Uruguay, the Colonia region, being a little cooler than Canelones, seems to be the best site for Pinot Noir to date: Finca Narbona and Los Cerros de San Juan show potential.

QUEBRANTA. *See p. 37*

SANGIOVESE

> *Grown in:* Argentina (4,970 acres), Chile (270 acres), Brazil (60 acres)
>
> *Distinctive in:* Very rarely distinctive

Italy's most widely planted red grape has spread to only a few other places in the world. One is the United States; another is Argentina, which has the largest holdings of Sangiovese after Italy. Alas, little of it is of memorable quality, and most ends up in simple red or rosé blends. Because of Brazil's historical connections to Italy, it is also planted there, though not in significant quantities.

Despite the emphasis on quantity over quality, the grape can make nice wine in Argentina. Good varietal examples of Sangiovese are produced by Finca La Luz (made from Uco Valley fruit) and Benegas (using Maipú fruit). Both are serious and packed with red cherry, spice, and notes of tarry mineral. Alas, those are the only notable examples. The rest of the Sangiovese crop is blended and consumed locally. The same is generally true in Chile and in Brazil, though there are always exceptions.

SYRAH

> *Grown in:* Argentina (31,650 acres), Chile (17,000 acres), Uruguay (215 acres), Brazil (50 acres), Peru, Bolivia
>
> *Distinctive in:* Chile (Aconcagua: Casablanca, San Antonio; Coquimbo: Elquí, Limarí; Central Valley: Rapel [Cachapoal]), Argentina (Mendoza: Luján de Cuyo, Uco Valley; San Juan: Pedernal Valley)

Some of the best examples of Syrah I have tasted in recent years have come from Chile (mostly from cool-climate areas), though there are some excellent examples in Argentina and a few each in Brazil and Uruguay. With a climatic range as great as Australia's, Chile produces some wines that are reminiscent of great northern Rhône examples, like Côte-Rôtie or Hermitage, while others are fleshier, evocative of Shiraz from Australia's McLaren Vale or Barossa Valley.

While Syrah is planted all over Argentina, I was especially impressed with wines from San Juan's Pedernal Valley, specifically from Las Moras. They bridge the gap between Old and New World styles admirably and clearly demonstrate the capabilities of the zone, a high-altitude part of San Juan, which is renowned for Syrah in a more

Shiraz-like interpretation. Terrific Syrah is made from Luján de Cuyo to the Uco Valley. These wines lean toward ripe blackberry and black cherry, with weight, spice, and varying amounts of oak. Memorable bottles include offerings from Bressia and Mauricio Lorca, and Laura Catena's "Luca."

Across the Andes, Syrah is fast becoming Chile's most memorable red grape. Wines from central Maipo are fat, straightforward, fleshy, and soft—reminiscent of southeast Australian Shiraz—while other regional styles are also emerging. Examples from cooler-climate Limarí exude pepper and tea leaf, with notes of herbs and blue fruit. The wines of Leyda and coastal Casablanca are smoky, granitic, and similar to Rhône wines (with green- and black-peppercorn notes). In Elquí, they have a distinct graphite note. To experience Chile's preeminence in Syrah, look to Falernia, Casas del Bosque, Undurraga's "T.H." series, Tabalí, Tamaya, GEO Wines, and De Martino. The potential of Uruguay's Syrahs can be tasted in Reinaldo De Lucca's bottling from the El Colorado area of Canelones: spicy, slightly feral, and framed with notes of sarsaparilla and fresh herbs.

TANNAT

Grown in: Uruguay (4,485 acres), Argentina (1,740 acres), Brazil (1,060 acres), Chile (10 acres), Peru

Distinctive in: Uruguay (Colonia, Canelones, Maldonado)

For me, the best Tannat anywhere in the world comes from Uruguay. The grape originated in Madiran in France and the Basque country around Iroléguy and Béarn. Named for its tannins, the grape can be bitter and astringent. But a combination of correct clones, focused vineyard practices, and well-managed vinification can tame the savage beast.

Tannat has a long history in Uruguay and is unmistakably the country's leading grape, accounting for three times more plantings than the second-place red variety, Merlot. Most producers will tell you that quality production began in the mid-2000s, with special attention to tannin management. Given the variety's longer growing cycle, it needs to be left on the vine as long as possible; growers must allow for some shriveling or raisining, but not too much. The timing of flowering and fruit formation is critical, as is tactical leaf plucking at just the right moment. Tannat quality seems to depend on three factors: tannin management in the vineyards, dropping yields substantially to maximize the potential of the fruit and balance tannins and greenness; picking as late as possible to ensure crunchy brown seeds with no green bitterness; and cold pre-fermentation maceration, which helps extract maximum flavor from the skins without astringency. Successful wines express ripe red berry, black fig, candied fruit, marmalade, and nuances of wood smoke, black pepper, and spice. Styles can resemble a range of wines, from port to Beaujolais. Cooler Maldonado and Colonia wines are a bit leaner in profile though still ripe and elegant. Those from Canelones differ as you move from

Progreso to Juanicó, from Colorado to Las Violetas, and don't yet carry specific flavor markers. As a rule, wines from these areas are more succulent than those from Colonia or Maldonado. A few producers to count on consistently are Irurtia and Narbona in Colonia; Pizzorno, Pisano, Carrau, and Bouza in Canelones; and Alto de la Ballena and Bodega Garzón in Maldonado. Others are noted later in the book.

TEMPRANILLO/TINTA RORIZ

Grown in: Argentina (15,130 acres), Chile (130 acres), Brazil (30 acres), Uruguay

Distinctive in: Argentina (Mendoza: Uco Valley, Eastern Mendoza)

In South America, the variety responsible for Spain's most famous red wines is of consequence only in Argentina, where it is planted abundantly and makes tasty, albeit simple wines. Almost all of the plantings are destined for blends rather than solo bottlings, but these blends always maintain the grape's characteristic red cherry and *balsamico* notes—a complex set of savory flavors that range from marjoram and mint to cinnamon, curry masala, and tobacco leaf. Tempranillo's personality on the rest of the continent is similar.

In Argentina, most Tempranillo is blended with either Bonarda or Malbec. But there are exceptions, driven by three producers: O. Fournier in the Uco Valley, whose estate-grown fruit makes for an exceptional wine; Familia Zuccardi, whose Santa Rosa vineyard source in Eastern Mendoza is splendid (yielding wines slightly less sharp than Tempranillos can often be); and Altocedro, with plantings of old vines in La Consulta (Uco Valley). These champions have coaxed out wines that show raspberry and red or black cherry, cinnamon, pink ginger, spice box, and tobacco leaf. I was told about a remarkable Tempranillo from Chilecito (La Rioja) but have not had the opportunity to taste it.

TOURIGA NACIONAL

Grown in: Brazil (90 acres), Argentina (40 acres), Uruguay

Distinctive in: Brazil (Rio Grande do Sul: Campanha Gaúcha, Serra do Sudeste [Encruzilhada do Sul])

Given Brazil's connections to Portugal, it's logical that this is the country with the most prominent plantings of this Portuguese variety. Touriga is grown in a few locations within the state of Rio Grande do Sul but appears to be best adapted to Campanha Gaúcha at the Uruguayan frontier, which is drier than regions farther north. Over the border, Touriga shows equally well in the Uruguayan state of Rivera. As in Portugal, it is used primarily in *cortes* with its traditional partner grapes (Touriga Franca, Bastardo, and Tempranillo) in addition to Tannat and Cabernet Sauvignon.

Although this grape has not yet spread far beyond Brazil, I believe it will be more prominent in South America in the future. The warmer and drier Campanha region is a prime spot for the variety, as Miolo's "Castas Portugeisas" demonstrates. This success has prompted other local growers to expand their plantings. Farther north, there is excitement in Encruzilhada do Sul over Lidio Carraro's noteworthy "Elos" blend. Dal Pizzol and Pizzato do a great job with bottling the grape as a varietal wine: their wines are extracted and dense, packed with the typical flavors of blackberry, bitter chocolate, lavender, and gingerbread.

UVINA. *See p. 38*
UVINA. *See p. 38*

CRIOLLAS

Criolla (meaning "creole") is a name used to refer to several grape varieties from Spain, including Criolla Grande, Criolla Chica (also known as Pais or Mission), and all three Argentine Torrontés varieties (Torrontés Mendocino, Torrontés Riojano, and Torrontés Sanjuanino), among others. These grapes came over with the conquistadores and Spanish settlers in the sixteenth century.

CEREZA

Grown in: Argentina (72,125 acres)

Distinctive in: Not very distinctive

Cereza, the dominant criolla variety, is the second most widely planted grape in Argentina. This grape, whose name means "cherry" in Spanish, is actually a cross between Pais (Criolla Chica) and Muscat of Alexandria. Known for its productivity, the grape makes mostly easy-drinking rosé and white wines.

High-quality Cereza is oxymoronic. Its sheer volume in Argentina, however, can't be ignored. With essentially as much land dedicated to this grape as there is to Malbec (almost 15 percent of all cultivation), Cereza is distinctively Argentinean and the driving force in basic table wine.

CRIOLLA GRANDE

Grown in: Argentina (42,200 acres)

Distinctive in: Not very distinctive

The fourth most widely planted grape in Argentina (mostly in Mendoza) is, like Cereza, the source of volumes of pink or rosé and white wines that are at best everyday quaffers and at worst innocuous. This variety is on the decline, more so than any of the other Criolla varieties.

MOLLAR

Grown in: Peru (acreage not available)

Distinctive in: Not distinctive

The more precise name for this grape is Negramoll; it is native to Andalusia in south-western Spain. Mollar is found exclusively in Peru, where it is used solely for the production of pisco, the national beverage. It was recently proved to be the offspring of Quebranta, with which it is usually interplanted, a grape prized for its ripeness and sweetness. It is rare to discover vineyards dedicated specifically to the cultivation of Mollar, and so it is equally rare to find a pisco made exclusively with Mollar grapes.

PAIS (CRIOLLA CHICA, MISSION, LISTÁN PRIETO)

Grown in: Chile (14,450 acres), Argentina (1,050 acres)

Distinctive in: Chile (Central Valley: Maule)

Originating in Spain's Castilla–La Mancha long ago, this variety is known as Pais in Chile and Criolla Chica in Argentina. In the United States, we know it as the Mission grape. Pais was brought from the Mexican territory of New Iberia in the mid-sixteenth century and was first planted in Peru before being cultivated farther south, after invasions of Argentina and Chile. Until 2008, it was the second most widely planted grape in Chile.

With close to three hundred years of history, Pais was the backbone of Chile's wine industry until eyes began wandering to France in the 1800s. With the recent emphasis on improving quality, the grape has fallen out of favor. There is a movement, especially in Maule, to preserve Pais and treat it with more respect in the vineyard to produce lower yields. However, this undertaking is more sentimental than it is effective in yielding brilliant bottles. Pais makes wines with, shall we say, unique savory and meatier flavors. It is usually vinified rustically into a light to medium cherry-red, orange, or light pink wine with biting acidity; this quality enables it to be made into tasty sparkling wine, as exemplified by Miguel Torres's Santa Digna "Estelado."

PEDRO GIMÉNEZ/PEDRO XIMÉNEZ

Grown in: Argentina (29,970 acres of Pedro Giménez and 20 acres of Pedro Ximénez), Chile (870 acres of Pedro Ximénez)

Distinctive in: Rarely distinctive anywhere

Pedro Giménez is a Criolla grape found exclusively in Argentina. Pedro Ximénez (note the slightly different spelling) is another immigrant grape from Andalusia, where it plays a major role in the production of sherry. Long the primary Argentinean source of white table wine (mainly supermarket box wines and large-format bottles), and still the seventh most widely planted variety there, Pedro Giménez is now on the decline. Pedro Ximénez is a key grape in the production of pisco in Chile, where more pisco is produced today than in its native Peru.

QUEBRANTA

Grown in: Peru (Ica Valley)

Distinctive in: Peru (Ica Valley)

Best known for its role in Peruvian pisco, this red-skinned grape is actually a cross between Pais and Mollar. It is found only in Peru.

In Peru there are five different pisco-growing regions, with forty-two valleys, eight approved grapes, and close to five hundred producers. Of the eight permitted varieties, four (Torontel, Italia, Albilla, and Muscat [Moscatel]) are considered aromatic grapes. The nonaromatics include Quebranta, Negra Peruana (known as Pais in Chile), Uvina, and Mollar. Opinions vary as to whether Quebranta, the most popular grape for pisco, is neutral or whether it adds flavors, which I have seen described as ranging from pepper and smoke to nuts and caramel. I suspect that the answer depends on whether the grape is used to make a pisco *puro* (monovarietal) or *acholado* (a blend).

TORRONTÉS

Grown in: Argentina (Riojano: 19,000 acres, Juanino: 5,050 acres, Mendocino: 1,760 acres), Uruguay (20 acres, unspecified)

Distinctive in: Argentina (Salta: Cafayate Valley, Calchaquies Valley; La Rioja: Famatina Valley; Mendoza)

Argentina's signature white variety is actually three different grapes—not, as is often suggested, merely different selections of Torrontés. Riojano, the most widely planted by far, is found in Salta and La Rioja and is a cross between Criolla Chica (Pais) and Muscat of Alexandria. Juanino is a white grape associated, as the name suggests, with the state of San Juan. Although it is of the same parentage as Riojano, it was genetically profiled in 1989 as being different.[*] Mendocino was documented as a separate variety in the same studies; it is associated specifically with Mendoza.

Torrontés took off in the 1980s during the white-wine boom, though it wasn't valued until the late 1990s. Its flavors suggest a meeting of Viognier, Gewürztraminer, and a

---

[*] See discussion in Jancis Robinson, Julia Harding, and José Vouillamoz, *Wine Grapes* (New York: Harper Collins, 2012), pp. 1066–67.

Hawaiian flower market. Winemaker Susana Balbo claims that the higher the altitude, the better the fruit. As the variety is quite productive (and often trained on pergolas), significant efforts are being made to reduce yields to improve quality. The consensus is that of the three grapes, Riojano, found in both Salta (the epicenter of Torrontés production) and La Rioja, shows the best balance between acidity and flavor and has the most persistence and length. Best consumed soon after vintage and as fresh as possible, Torrontés is a favorite of the U.S. market, which is the leading importer (it imports five times more than the second-place United Kingdom). Exploding with flavors that range from the tropical (passionfruit and mango) to stone (apricot, nectarine, and peach), it has signature floral nuances of honeysuckle, freesia, and plumeria. Dependable producers include Susan Balbo with her "Crios," Finca Las Nubes, Colomé, Alta Vista, Trapiche, and Etchart.

UVINA

> *Grown in:* Peru (acreage not available)
> *Distinctive in:* Peru (Lunahuaná Valley)

Anecdotally claimed to be related to Muscat, this traditional variety hails from Peru's Lunahuaná Valley. Used for pisco, it's found today in the valleys of Lunahuaná, Pacaran and Zuñiga, and Cañete-Lima.

Uvina, characterized by its small bluish-black berries and large clusters, adds structure and a signature green-olive note to pisco. It's one of the four nonaromatic varieties approved for the production of pisco.

3

ARGENTINA

Planted acreage: 546,000

World rank in acreage: 9

Number of wineries: 1,320 (about 370 making fine wine)

Per capita consumption (liters)/world ranking) 26/18

Leading white grapes: Cereza, Chardonnay, Sauvignon Blanc, Torrontés

Leading red grapes: Bonarda, Cabernet Sauvignon, Criolla Grande, Malbec

Memorable recent vintages: 2002, 2004, 2005, 2006, 2009, 2010, 2011, 2012, 2013

View from the vineyard of the Salentein Winery in the Uco Valley in Mendoza. Photo by Matt Wilson.

INTRODUCTION

Six years ago, anxiously arriving in Mendoza City for the first time, I asked the driver two questions. First, why were there empty plastic bottles standing on the roofs and hoods of so many cars? Answer: to inform passersby that the cars were for sale. Second, why is Argentina, "land of silver," called Argentina? Are there significant silver mines? Is it named for Argentina's enormous river of same name (the Río de la Plata)? Answer: that's a good question, but there's no universally accepted response. I replied that it might be appropriate to call the country Viñotina, "land of wine," given its importance to Argentine culture. He smiled. When I added that no other country in the world had chosen wine as its national beverage by executive decree (as Argentina did in November 2010), he smiled even more.

Wine is vital to Argentina. The country is the world's seventh-largest wine consumer by volume, the largest in the Southern Hemisphere; and, after the United States, it is the second-thirstiest nation in the New World (again based on wine volume consumed). Though its per capita wine consumption has plummeted over the years and is now rivaled by Uruguay's, Argentina's significantly larger population more than compensates. Moreover, Argentina's export market remains strong: the country is the third-largest wine exporter to the United States, a trade that was worth $273.7 million dollars in 2013, according to U.S. Customs figures.

Argentina's success is remarkable considering that as recently as 2006 it didn't even rate its own section in Tom Stevenson's annual *Wine Report*. Only five years later, Argentina was awarded more international medals in the 2011 *Decanter* magazine wine awards than any country since the competition began in 2004.

Argentineans are passionate people, a likely result of the fact that 70 percent of them have Italian blood somewhere in their veins. This also gives them a time-honored independent streak. An old adage about Argentineans is that they speak Spanish, dress like Italians, behave like the French, and think they are English. Like any stereotype, this observation is only partially valid, but it points up the country's spirit of fervent individuality and its history of diverse waves of immigration. Many vivid images of Argentina enhance its allure: the sensual tango, the awesome beauty of the Andes and the pampas, a celebrated epicurean culture, and the vibrant metropolis of Buenos Aires, the so-called Paris of South America. And then there's the wine.

Spanning one thousand miles from north to south, from Jujuy's Maimará to Patagonia's El Hoyo de Epuyén, Argentina's wine regions possess a unique combination of attributes. High-altitude vineyards (also found in Peru and Bolivia) and a dry desert climate create vast potential for wine. Lower rainfall equates to less humidity, which means less rot, and to environments that are relatively free of pests. The main worries for growers are not bunch decay, botrytis, or insects but rather hail, frost, and strong winds that can inhibit fruit set. Every five years or so, a spring frost can destroy 30 percent or more of the vintage in the Uco Valley. In the greater region of Mendoza, some 31,400 acres are protected from hail with nets.

Argentina is not entirely free of pests. Like Chile, it is fortunate to be free of phylloxera, the root louse that has caused periodic vineyard devastation worldwide (though Argentineans don't trumpet this fact as boldly as the Chileans do). Flood irrigation prevents the root louse from completing its full life cycle above ground, resulting in one

WHERE'S THE BEEF?

Although Argentineans are known as the world's most enthusiastic consumers of beef, people eat roughly as much beef in Southern Brazil and Uruguay as they do in Argentina. The pampas, the elevated grass plains of South America, cover three hundred thousand square miles from the Atlantic Ocean to the Andes (between 34° and 30° south and 57° and 63° west). While situated primarily in Argentina (incorporating the provinces of Buenos Aires, La Pampa, Santa Fe, Entre Ríos, Córdoba, and Chubut), this terrain extends into Uruguay and Brazil's Campanha Gaúcha. The word *pampa* comes from the Quechua word for "level plain." On these high plains, it is said, there are 190 different types of grass. The cows that feed on it produce beef that is lean and exceptionally flavorful (when properly cooked), and they are bred specifically for eating, not dairy farming. Hung dry and on the bone for three weeks, Argentinean grass-fed beef is prepared to perfection by talented *asadors*, grill masters extraordinaire, who may be either restaurant professionals or gifted amateurs hosting a weekend *asado*. Served with chimichurri—a sauce of parsley, garlic, and chili—Argentinean beef is a carnivore's dream come true.

Grilled beef in Argentina is very different from steak at a North American backyard barbecue. The meat is not marinated or sauced but simply seasoned with salt and pepper. The coals are spread in a rectangular shape (leaving space in the middle), creating a gentle, even heat that surrounds the meat, and the beef is cooked very slowly for at least an hour, until it emits a gentle but constant sizzling sound. It's said that the slow cooking allows the fats to cook through the meat, leaving it lean and tasty.

less vineyard concern. Nematodes, on the other hand, can be problematic: damage is mitigated by vineyard work. And although climate change means that both Argentina and Chile will face long-term agricultural problems, for now Andean snowmelt provides an ample source of pristine water. Argentina's advantages are best summarized by Steven Spurrier, who writes that it "has no rivals for the ripe flavors, reliability and value for money of its reds and possibly none [in the New World] when it comes to dramatic improvements in the future."[*]

Like people elsewhere in South America, Argentineans typically drink "common" (table) wine as opposed to "fine" wine. The foundation of everyday wines in Argentina is criolla varieties, namely Cereza, Criolla Grande, and Pedro Giménez. *Vinifera* grapes form the backbone of quality production, led by Malbec and Cabernet Sauvignon. A couple of other *vinifera* varieties—notably Bonarda and Tempranillo—bridge the two

[*] Steven Spurrier, "Argentinian Red Wines," *Decanter*, June 4, 2008; www.decanter.com/people-and -places/wine-articles/485786/argentinian-red-wines#LwzPsWt6O2bhtPEI.99.

genres, often being blended with other *vinifera* grapes or with criollas to produce modestly priced fine wines.

Argentineans consumed far more wine in the 1970s than they do today. The traditional Argentine palate was "old Spanish," leaning toward oxidative styles and wines based on the aforementioned less interesting criollas. Younger people and newly discerning drinkers have eschewed these older styles of wine and also consume greater quantities of other beverages, such as beer and soda. These changes, along with dramatically declining consumption of table wines, internationalization, and a new and vibrant export market, have led to the emergence of an export-focused premium wine industry.

Contemporary restaurants worldwide have embraced the new offerings from Argentina, and wine lovers from overseas are arriving in increasing numbers to discover the wines for themselves. Close to 170 wineries make up a busy wine-tasting circuit in Mendoza: in 2011, it hosted more than 500,000 tourists, an increase of 68 percent over 2009.[*]

Argentineans consume 149 pounds of beef per person per year, more than any other country in the world. Why? Maybe it's because they need something to accompany all that delicious Malbec!

HISTORY

Argentina's modern wine industry was arguably born on May 25, 1885. That day marked the opening of the railroad line between Mendoza, Argentina's principal wine-producing area, and Buenos Aires, the country's largest consumer market. Although wine had been produced in Mendoza since the mid-sixteenth century, transport on the backs of mules to the urban centers was expensive and slow, and much of the wine spoiled in transit. The railroad reduced delivery time to three days, bringing wine to market in better condition and at lower prices. The floodgates opened.

Mendoza's provincial government had begun preparing for this birth some years before, instituting large irrigation projects for the vineyards, establishing tax exemptions for the planting of vineyards in 1881, and recruiting Italian and Spanish immigrants with farming experience. Argentina's welcoming attitude to immigrants was summed up in the motto *poblar es gobernar* (to populate is to govern). That, and its promise of peace and prosperity, made it an attractive destination. By 1910, almost 70 percent of the wineries and 83 percent of production were in the hands of immigrants with backgrounds in European wine culture. In 1852 the French botanist Michel Aimé Pouget, under contract to the Argentinean government, arrived from Santiago and

---

[*] Mindy Joyce, "Wine Tourism Is Increasingly Being Sought Out by Travelers, but Is the Wine Industry Ready?" http://mindyjoyce.com/2011/07/10/wine-tourism-is-increasingly-being-sought-out-by-travelers-but-is-the-wine-industry-ready/.

introduced the first French *vinifera* varieties to Mendoza. In 1885, Tiburcio Benegas, a pioneer in Argentinean viticulture, was convinced that these new grapes needed to be kept pure and bred separately, an uncommon practice at the time. His Trapiche wine company established the framework for quality growing in Argentina.

By the 1920s, Argentineans were drinking sixty-two liters of wine apiece per year; together the Italian and Spanish immigrants consumed 80 percent of all the wine produced. Wine was considered essential to the daily diet, an important source of calories and an excellent substitute for nonpotable water. After bread and meat, it occupied third place in the family shopping basket. For these immigrants, largely poor and male, taste was not a significant consideration. Their main concerns were availability and price.

Grape production increased at a steady rate of 2–4 percent annually through the 1950s. Then it boomed. In 1972–73 alone, the total amount of land in grape production expanded by 10.5 percent, and by 1977 it peaked at 865,000 acres. Consumption increased to match, reaching ninety liters per person by 1977 (ninety-nine liters in Buenos Aires proper).[*]

From the late 1970s and into the early 1980s, however, the wine industry went into a decline as a result of the burden of government debt, unemployment, and inflation incurred by Argentina's military government. In 1989, with a declining gross domestic product and inflation in excess of 5,000 percent, the economy collapsed. The economic situation drastically decreased wine consumption and led to a major pull-up of older unproductive vineyards, including much of the older-vine Malbec, locally called "French." With this reconversion of vineyards, cost-conscious growers focused on the prolific criolla varieties and higher-volume *vinifera* grapes like Bonarda. A sudden consumer craze for white wines led growers to concentrate on high-volume, lower-quality whites, and many producers matter-of-factly made white wine with red grapes. Production of grape concentrates also increased sharply: in the 1980s, 25 percent of all grapes planted were used mainly for concentrate. And a cultural shift meant that Argentineans were beginning to work longer hours, obliging them to forgo long, wine-accompanied lunches and siestas.

Starting in 1989, President Carlos Menem took the country in a different direction. Among the gamut of new economic policies he introduced, the most significant was the 1992–93 pegging of the peso to the U.S. dollar. This put a stop to fears of inflation. The economy stabilized, the fiscal deficit plummeted, and inflation came to a halt. Also, as Argentineans now effectively held dollars, they could afford to travel abroad and buy imported goods, and they had access to credit at low rates for the first time in decades. As a result, quality of life skyrocketed. These new policies also encouraged foreign investment in the Argentinean wine industry. Incorporating international standards

* Steve Stein, "Culture and Identity: Transformations in Argentine Wine, 1880–2011," unpublished paper. http://estructuraehistoria.unizar.es/gihea/documents/Stein.pdf, accessed January 15, 2014.

THE INTERNATIONAL ANGLE

A large number of Argentinean wineries work with international consultants or are foreign owned, more than anywhere else on the continent, although the leading wine companies—including Catena, Luigi Bosca, Zuccardi, and Peñaflor—are owned by Argentineans. One reason for this reliance on consultants is that many current owners, Argentinean and otherwise, do not come from the wine industry. In the early 2000s, even the established wineries that were seeking to improve quality hired consultants, and the wineries simply did what the consultants told them to do. As the industry has matured and winemakers have learned their trade, consultants are now more often invited to give an opinion, on the basis of their global expertise and more objective view of the wines, and a discussion ensues.

Another fact of life is that the services of a famous consultant add value to the wine. A winery advised by an international luminary such as Paul Hobbs or Michel Rolland can put a higher price tag on its bottles. However, Argentinean wineries are increasingly turning to "local" consultants, such as Susana Balbo, Roberto de la Mota, Mauricio Lorca, Matías Michelini, Luis Barraud, and the Durigutti brothers Hector and Pablo, who are globally respected but more readily available.

All of the foreign consultants have put their money where their palates are, by owning or investing in wineries in Argentina: Paul Hobbs with Viña Cobos; Michel Rolland with his Michel Rolland collection of estates, Val de Flores, Mariflor, and Casarena; and Alberto Antonini with Altos Las Hormigas. As Paul Hobbs remarked in a 2009 presentation on investing in the wine industry, "The decision to invest in Argentina was one of the most straight-forward. . . . Like an untapped vein of liquid viticulture gold, there is an extraordinary proportion-potential unknown and unrealized. . . . It is a prospector's dream come true."

opened the minds of producers at the established bodegas and led to the creation of new, foreign-owned operations.

In January 2002, the peso-to-dollar peg was abandoned. In a matter of days, the peso lost significant value in the currency market. Such a change would normally lead to an increase in exports, but since the country's economy was structured for an import-heavy market, export income was slow to expand. In consequence, the value of the peso continued to drop, bottoming out at approximately 3.15 pesos to the dollar. Export sluggishness also reduced tariff revenues and government savings, which in turn caused inflation to rise significantly once again, reaching 20.2 percent in April 2002. In response, the government began to aggressively promote exports with subsidies and improved access to credit for export-related industries. With increasing soy prices, Argentine exports eventually began to gain a footing in international markets, and wine has benefited from these developments.

ARGENTINEAN GAME CHANGERS

SUSANA BALBO

Today in Argentina, only one in ten winemakers is female—and that's a big jump from past numbers! Susana Balbo stands out not only for her gender but also for her incredible skill and experience. Her original intent was to become a nuclear physicist, but, as her parents were against this career choice, she ended up studying enology in Mendoza, becoming Argentina's first graduate female winemaker in 1981. Over thirty years later, she is not only one of the country's most experienced winemakers but also among the first to be hired as an international consultant: she has made wine in Australia, California, Chile, France, Italy, South Africa, and Spain, and she faithfully spends one month a year in a different wine region of the world, studying with local winemakers and growers. A former president of Wines of Argentina, Balbo has her own label, Dominio del Plata, which is among the country's most successful.

NICOLÁS CATENA

Nicolás Catena, the undisputed patriarch of Argentinean fine wine and a third-generation vintner, is considered to be the godfather of Malbec. This and other accomplishments led to his being named *Decanter* magazine's "Man of the Year" in 2009, a first for South America. Nicolás's grandfather planted his first Malbec vineyard in Mendoza in 1902. In the early 1980s, Nicolás took an inspired visit to California, where he was a visiting professor of economics at the University of California, Berkeley, and spent many weekends in the Napa Valley learning about premium wine. Upon returning to Argentina, he made a push toward quality. He embarked on soil studies, vineyard experiments (climaxing in the now-famous trial of 145 selections of Malbec propagated in his Angélica vineyard to identify the best choice for Mendoza), and engaged consultants like Paul Hobbs, Attilio Pagli, and Jacques Lurton. A pioneer in the use of extended macerations and small, new French oak barrels (circa 1990), Nicolás was also the first to engage in the export market. His efforts were recognized when the 2004 "Nicolás Catena Zapata" became the first Argentine wine to score more than 98 points in Robert Parker's *Wine Advocate*.

ROBERTO DE LA MOTA

One of the most knowledgeable and respected wine experts in Argentina, Roberto was born into the business: his father, Raúl de la Mota, produced the iconic 1977 "Cavas de Weinert" Malbec, considered the country's most historically significant fine wine. Raúl himself was influenced by Bordeaux's eminent Professor Émile Peynaud. In the 1980s, Roberto was instrumental in bringing in new varieties like Viognier, Petit Verdot, and Cabernet Franc from France and worked with vineyards to adapt to new conditions and techniques, such as shifting from traditional flood irrigation to modern drip systems. In 1994, Roberto went to work for Moët & Chandon's Argentinean outpost developing the new Terrazas de los Andes winery, and then, with Château Cheval Blanc's Pierre Lurton, he helped to create Cheval des Andes before leaving to set up his own winery, Mendel. He still consults for many bodegas and is perennially sought after for his advice.

JULIO VIOLA

A visionary real estate dealer in Neuquén, Julio Viola bought 7,900 acres of desert just north of the region's namesake city in 1996 with a government grant. After going to Israel to study drip irrigation, he installed an irrigation system for the land that included 3,800 miles of piping and seven pumping stations. Then, beginning in 1999, he planted vines. Keeping some 1,075 acres for himself, he sold off the other properties and used the proceeds to start his own winery, Bodega del Fin del Mundo. Single-handedly he created the San Patricio del Chañar wine region, which has fast become known for its high quality and leading Pinot Noir. Admittedly a novice when he started, Viola is very hands-on and actively engaged in this vibrant region, which now boasts seven wineries. In part because of the area's remoteness, he opened a very chic restaurant in Buenos Aires's Palermo neighborhood to serve as a showcase for his wines.

REGIONS AND WINERIES

Argentina's immense wine region extends over a thousand miles of latitude, from the northern province of Salta to the southern region of Patagonia. The principal winemaking areas are in the northwestern provinces (Salta, Catamarca, and Tucumán), the central provinces of Cuyo (La Rioja, San Juan, and Mendoza), and the southern provinces of Patagonia (Neuquén, La Pampa, Río Negro, and Chubut). The provinces of Mendoza and San Juan account for about 90 percent of the total area planted. In 2010, there were some 24,700 registered vineyards, of which 16,300 were in Mendoza.[*] Red grapes represent approximately 80 percent of fine-wine production, driven by Malbec, Bonarda, and Cabernet Sauvignon; the leading white varieties are Torrontés and Chardonnay. The cultivation of Malbec, Argentina's leading variety, is three times that of Bonarda, with 55 percent of the Malbec plantings situated in Mendoza.

Argentina has three educational institutes for wine and agronomy, which attract students from all over South America. The leading program is widely thought to be at the Don Bosco Enological University in Mendoza, though excellent instruction is also offered by the enological departments of the University of Mendoza and the Agricultural College of the National University of Cuyo, also in Mendoza.

[*] USDA Foreign Agricultural Service, Global Agricultural Information Network report, "Argentina: Wine Annual, 2012," http://gain.fas.usda.gov/Recent%20GAIN%20Publications/Wine%20Annual_Buenas%20Aires_Argentina_3-2-2012.pdf, p. 3.

Map 2. Argentina's wine regions

HIGH-ALTITUDE WINES

We know what happened when Icarus flew too near the sun: the wax in his wings melted, the feathers fell off, and he crashed in punishment for his hubris. Grapes in Argentina, by contrast, adapt and thrive at high altitudes. They adapt to the higher ultraviolet exposure by growing thicker skins and developing deeper pigmentation. Astute viticulturists know how to manage leaf canopies to provide optimal protection from ultraviolet radiation and wind. Higher vineyards are cooler: there's an average drop in temperature of one degree Fahrenheit for every 328 feet of elevation. This gradual decline is accompanied by increases in thermal amplitude (the difference between day and nighttime temperatures), a factor that extends the growing season and benefits the fruit. Daytime warmth develops sugars, promotes ripening, and softens tannins; the evening's chill preserves the grape's acidity and balanced structure. Coupled with the dry climate and carefully timed irrigation, these conditions are close to perfect for growing fruit.

In 1999, the Instituto Nacional de Vitivinicultura (INV) passed appellation-of-origin laws governing the use of names of all wine regions in Argentina. In order for a wine to be labeled with a specific appellation, described by the law as *indicación geográfica* (IG), 100 percent of the grapes must come from the appellation listed on the label. In order for a wine to be associated with a specific province (with the designation IP, or *indicación provincial*), such as Mendoza or Salta, 85 percent of the grapes must originate from that province.

What makes the Argentina law a bit confusing is the fact that some region names are also the trademarks of wineries. For example, Uco Valley's appellation of Altamira is home to vineyards owned by Catena Zapata, Familia Zuccardi, Chandon Argentina, and Achaval Ferrer, as well as several smaller family producers, such as Pizzella and Alegría. However, Achaval Ferrer and Altamira Wines, a United States company, together own the trademark to the name Altamira and have thus far refused to let the region's producers use it on wine labels. This situation is analogous to Robert Mondavi's claiming the exclusive right to list Napa Valley on a wine label! Reform or compromise is required. A successful example of resolution is Finca La Celia's ownership of the trademark *La Consulta*. It has allowed other producers in the La Consulta district to use the trademark as an appellation, with some cosmetic restrictions. Resolving these many disputes has become a pet project of the tireless Laura Catena, Nicolás's daughter and overseer of Catena Zapata's global business.

Another labeling detail worth knowing about is the use of the designations *Reserva* and *Gran Reserva*. Reserva wines require a minimum of six months of aging for white wines and one year for reds; the Gran Reserva designation requires at least one year of aging for whites and two for reds.

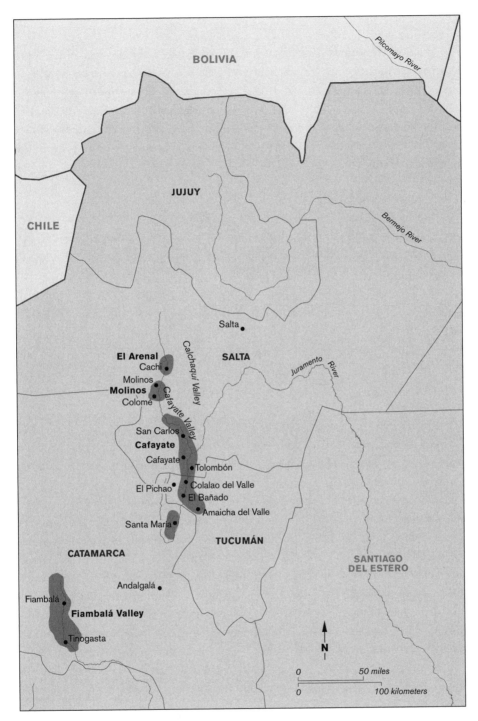

BOLIVIA

Pilcomayo River

JUJUY

CHILE

Bermejo River

Salta •

El Arenal •
Cachi •

SALTA

Calchaquí Valley

Juramento River

Molinos •
Molinos
Colomé •

Cafayate Valley

San Carlos •
Cafayate
Cafayate •

Tolombón •

El Pichao •

Colalao del Valle •
El Bañado •
Amaicha del Valle •

Santa María •

TUCUMÁN

CATAMARCA

SANTIAGO
DEL ESTERO

Andalgalá •

Fiambalá •
Fiambalá Valley

Tinogasta •

N

| 0 | 50 miles |
| 0 | 100 kilometers |

Map 3. Argentina: The northern region

JUJUY

This small, northernmost viticultural area of Argentina, home to some of the highest vineyards in the country, is not yet commercially established; virtually no wine is exported beyond the region's administrative boundaries. The extreme temperatures resulting from its tropical latitude (23° south) are moderated by its high altitude. Located near the borders of Chile and Bolivia, the administrative province of Jujuy sits almost entirely on the eastern slope of the Andes. The area is best known for its Torrontés, though its red wines are improving.

SALTA

Salta takes its name from the Aymara word *sagta*, meaning "beautiful one." The native Indian culture is strongly present in this region of northwest Argentina. Celebrated for agriculture, Salta is also known for tobacco, bananas, and sugar, indicating the diversity of its climate. Local food is a big deal here, and the region is replete with *peñas folclóricas*, eating houses where you can enjoy native food, local wine, music, and traditional dancing. The wine region is about three hours from the eponymous capital of the province, near the border with Catamarca. The spectacular drive, on steep and narrow mountain roads through the ever-changing canyons of the Cochas River, is not for the faint of heart. The highest vineyards are located more than nine thousand feet above sea level.

There are three wine areas: Cafayate (Valle de Cafayate), Molinos (Colomé), and El Arenal. The larger Cafayate Valley, and its subregion of the Calchaquí Valley (which it shares with Catamarca and Tucumán), is the main wine-producing district, accounting for 70 percent of Salta's 6,200 acres of vines. The average altitude of Cafayate is 5,500 feet, and the steep vineyard slopes are somewhat reminiscent of those in Germany's Mosel region. Despite the altitude, the grapes receive adequate warmth and sunlight: solar radiation is more intense in the thinner air, and the soils of Cafayate contain pulverized mica, which Alejandro Sejanovich, of Manos Negras and Anko, compares to a terrain of small mirrors reflecting the sunlight from the ground up to the vines. Cafayate's subarea of Yacochuya is also very good. The other prominent region is Molinos: it is the location of an unofficial district, Colomé, home to the well-known namesake winery and a few others. Finally, there is increasing buzz around Angastaco, in the San Carlos department, and Tolombón, south of Cafayate as you approach Tucumán.

Most of Salta's reputation is based on its celebrated Torrontés, which accounts for almost 40 percent of all grapes planted, but the excellent peppery Malbec and structured, mineral-scented Cabernet Sauvignon wines are well worth a mention. Fifty-five percent of Salta's grapes are red, and almost all are cultivated for fine wine, which can be overripe, almost roasted, without careful viticulture and harvesting. Because of Salta's remote northern location, fruit is either trucked down to Mendoza for processing or, more often, custom-crushed and vinified and then brought down to Mendoza, either bottled or ready for bottling.

RECOMMENDED PRODUCERS

Amalaya

Bodegas Etchart (*see also* Pernod Ricard)

Bodega Tacuil

Colomé

Domingo Molina

Finca Las Nubes

Michel Torino (*see also* Grupo Peñaflor)

San Pedro de Yacochuya

TUCUMÁN

Tucumán, in the northwest, is the most densely populated but geographically smallest province of Argentina. Due to private efforts and state subsidies, Tucumán's wine industry is improving. There are currently just a small number of wineries and a couple of dozen growers in Amaicha del Valle, Colalao del Valle, El Pichao, and El Bañado. Plantings are primarily red grape varieties: Malbec, Cabernet Sauvignon, and, to a lesser extent, Merlot and Tannat. White varieties include Torrontés and a little Sauvignon Blanc. Even if they aren't yet coming for the wine, visitors enjoy the rich history of the area and its food: the empanadas found in Tucumán are said to be among the best in all of Argentina, and dedicated foodies can take the "Ruta de la empanada Tucumana" (the Tucumán empanada road) by following a map of close to sixty empanada-tasting spots around the capital city and suburbs.

CATAMARCA

Known for its heat and dry climate, Catamarca is located in the foothills of the northern Andes, which separate it from the Atacama region of Chile. The principal wine regions, the Fiambalá and Tinogasta valleys, both run north to south between protective hillsides. The 5,700 acres planted are not acknowledged for export-quality wines and historically have been cultivated with criolla grapes, a full 50 percent being planted to Cereza. However, as with other Argentinean regions, quality is improving, and producers are making wines with commercial promise from Cabernet Sauvignon, Malbec, and Syrah.

The more important Fiambalá Valley, a subsection of the larger Tinogasta Valley, sits at roughly five thousand feet above sea level and at 27° south. Sheltered to the east and west by steep hillsides, the peaks of which create a rain shadow, Fiambalá depends on irrigation from snowmelt. And with only a limited area suitable for viticulture, the Fiambalá vineyards are all located within a single concentrated zone, running north to south along the valley. Other, smaller wine areas in Catamarca include Andalgalá and Santa María, the latter being a focus of recent investments.

Cabernet de los Andes (*see also* Luigi Bosca—Familia Arizu)

LA RIOJA

Do not confuse this La Rioja with the celebrated namesake appellation in northeast Spain (Rioja). Located in the Andean region in the west, at 5,600 feet above sea level, the Argentinean La Rioja is the oldest and the third-largest wine region in the country, with 17,000 planted acres. Being far hotter and drier than the more southerly regions (summer temperatures of 110°F are not uncommon), the wine-producing areas of La Rioja have been strictly delineated by their access to water. This has produced a scattered pattern of vineyards. The main production area is in Chilecito, about 118 miles northwest of the capital of La Rioja and encompassing the subregions of Nonogasta, Cuesta de Miranda, Villa Unión, and the Famatina Valley, the premier wine district. The Famatina Valley is home of La Rioja's La Riojana, the huge, forward-thinking co-op, with more than five hundred members, that spearheaded much of the research on Torrontés, the region's most important grape. The five most important varieties of the province are Torrontés Riojana (30.3 percent), Cabernet Sauvignon (11.7 percent), Syrah (8.8 percent), Bonarda (7.7 percent), and Malbec (7.4 percent), collectively representing 65.9 percent of the cultivation.

RECOMMENDED PRODUCERS

La Riojana

Valle de la Puerta

SAN JUAN

San Juan, along with La Rioja and Mendoza, forms part of the larger central wine region of Cuyo ("land of sand" in the native Huarpe dialect), which accounts for 90 percent of Argentina's wine production. San Juan, the country's second-largest wine region, accounts for about a third of the area of Cuyo, with 116,700 planted acres at altitudes ranging from 900 to 4,600 feet. In 2010, the province of San Juan claimed a full 21.1 percent of Argentina's total number of vineyards and 21.7 percent of the planted area.[*] It encompasses twenty unique wine-producing areas in the five main valleys (Tulum, accounting for 89 percent of San Juan's total production; Ullum and Zonda, 6 percent;

---

[*] USDA Foreign Agricultural Service, Global Agricultural Information Network report, "Argentina: Wine Annual, 2012," http://gain.fas.usda.gov/Recent%20GAIN%20Publications/Wine%20Annual_Buenas%20Aires_Argentina_3-2-2012.pdf, p. 3.

Pedernal, 4 percent; and Calingasta, 1 percent). Of the three hundred producers in the area, fewer than ten focus on export-quality wines: the region has always been the heart of high-volume production in Argentina, and change has come slowly. Irrigating the vast areas of grapes required the construction of some 1,240 miles of irrigation channels (the distance from Los Angeles to Seattle). The San Juan River feeds three of the main valleys—Tulum, Ullum, and Zonda. The Tulum Valley surrounds the capital city of San Juan, while the other two valleys are within reasonable reach: Ullum is some twelve and a half miles northeast of Tulum, on the north banks of the San Juan River, and Zonda is directly west. Pedernal and Calingasta, two newly prominent areas, are more remote.

The Tulum Valley is the largest wine-growing subregion. At a latitude of 31° south, viticulture is possible here because of favorable local topography and irrigation from the San Juan River. The average altitude of 2,200 feet above sea level moderates the temperatures. Soils here are sandy, with clay throughout the five subregions: San Martín, Sarmiento, 25 de Mayo, 9 de Julio, and Caucete. Tulum is best known for Chardonnay and Torrontés (Juanino), though Syrah from this area is gaining recognition.

The Zonda Valley derives its name from the prevailing wind, created by the country's mountainous topography, which brings warm, dry air sweeping down from the hillsides. Although the wind lessens the risk of vine disease and brings warmth to high-altitude zones, violent gusts can damage the vines. While not renowned for quality wines—most of its grapes until very recently were criollas—the Zonda Valley is developing a fine-wine savvy. The adjacent Ullum valley concentrates on table wines.

Pedernal, the highest wine-growing subregion (the average vineyard altitude here is over four thousand feet), is located in the south of the region some sixty miles north of Mendoza and roughly the same distance from the Chilean border. Viticulture started here in 1993, and the area has quickly become one of Argentina's most exciting new vineyard sites. Pedernal is a glacial valley, named for its dark, flat silex flint stones. In this very dry climate, with formidable thermal amplitude, water is supplied through drip irrigation from local underground aquifers. Pedernal wines have a distinct minerality and an almost saline character, reminiscent of the wines of Mendoza's Gualtallary. While Chardonnay, Malbec, and Cabernet Sauvignon are planted, the star is some of very best Syrah in all of Argentina.

RECOMMENDED PRODUCERS

Bodega Augusto Pulenta

Bodegas Etchart (*see also* Pernod
  Ricard)

Bodegas La Guarda

Callia

Graffigna (*see also* Pernod Ricard)

Las Moras (*see also* Grupo Peñaflor)

Mumm Argentina (*see* Pernod Ricard)

XumeK

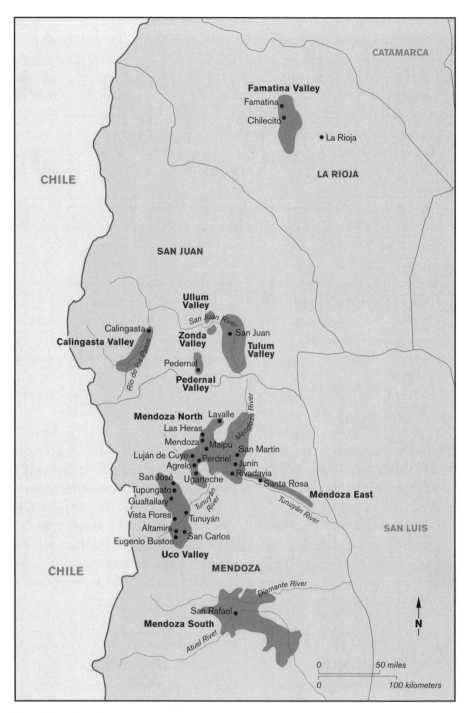

Map 4. Argentina: Cuyo

Agriculturally diverse Mendoza, a province as large as Spain, offers more than grapes. Eastern Mendoza specializes in onions, tomatoes, and carrots, and Tupungato is known for its walnuts. Flavorful peaches grow throughout much of the Uco Valley. And, after China, Mendoza is the second largest producer of garlic in the world. The region also accounts for 66 percent of all vineyards and 70.8 percent of all agricultural land in Argentina. Sitting squarely in the western rain shadow of the Andes, greater Mendoza's wine regions benefit from a dry climate, high average temperatures, and wide thermal amplitude. The average elevation of vineyards is around three thousand feet. Irrigation is facilitated by the region's various rivers, including the Mendoza River, and subterranean aquifers that are replenished annually by glacial melting. Drip irrigation (now used in about 15 percent of the vineyards) is gradually replacing flood irrigation. Warm, dry harvest periods result in grapes that are picked according to ripeness rather than weather patterns. Frost and hail are the most significant hazards to viticulture: hail-protection measures are used in 97 percent of the vineyards. Soil types vary across the region, but the majority of vines are planted on alluvial soils, locally called *franco*.

The primary five varieties of Mendoza are Malbec (17.2 percent), Cereza (10.8 percent), Criolla Grande (10.7 percent), Bonarda (9.9 percent), and Cabernet Sauvignon (8.2 percent). But for most wine lovers, Mendoza means Malbec. Malbec represents 14.2 percent of total plantings in Argentina, an increase of some 90 percent since 2000, and about 80 percent of that acreage is in Mendoza. Close to five hundred wineries in Argentina make Malbec, and about 40 percent of them export wine to the United States.

Malbec thrives in Mendoza for two main reasons. The first is that the region has very low annual rainfall (twelve inches compared to thirty in Bordeaux, Malbec's Old World home), and the rain falls mostly in the summer, not the winter. This timing promotes ripening and minimizes disease. It's said that there are 320 days a year of sunshine in Mendoza! Second, Mendoza's wide thermal amplitude promotes aromatic development (including Malbec's signature violet aroma) and softened (polymerized) tannins. The high altitude brings lower humidity and increased exposure to sunlight than the variety receives in France. Grape skins grow thicker at high altitudes; this is the grapes' way of protecting themselves against ultraviolet radiation damage. Thicker skins contain high quantities of pigment, tannins, and flavor compounds and, in wetter climates, help protect the fruit against rot.

Mendoza is divided into several zones. The Eastern Zone is home to the suburban wine-producing areas of Junín, Medrano, Rivadavia, San Martín, and Santa Rosa. This zone is marked by stony soils at higher altitudes and deeper soils in the more typical, lower spots. The area, which is irrigated by the Tunuyán River, is a source of vast quantities of everyday wines, both criolla and *vinifera*, and is most celebrated for its very good Bonarda. Northern Mendoza, which includes Las Heras and Lavalle, is also acknowledged for easy-drinking, everyday styles of wine.

The heart of the region, closest to the city of Mendoza, is often referred to as the Upper Zone and contains the oasis departments of Luján de Cuyo and Maipú, which produce many of the finest wines in the country. Together, these areas are about equal in size to Burgundy, encompassing around 67,000 acres.

Luján de Cuyo

About as large in cultivated area as the Napa Valley in California, Luján de Cuyo is Argentina's first official appellation, established in 1993. It encompasses 33,285 acres planted to different varieties: 44 percent Malbec, 18 percent Cabernet Sauvignon, 5 percent Merlot, 5 percent Chardonnay, and the rest made up by many other grapes. Hail can be problematic: growers typically lose 5 percent of their crop when hail pounds the vineyards.

Home to many of the best wineries in Mendoza, Luján de Cuyo is a tapestry of several subregions, some officially recognized and others still awaiting formal recognition. Unofficial Ugarteche, at about three thousand feet above sea level, is noteworthy for sandy and loamy soils irrigated by the Mendoza River. It produces wines typified by Luján's characteristic black fruit intensity and Ugarteche's classic notes of *garrigue* (the aromatic scrub vegetation of the coastal Mediterranean hills). Nearby Perdriel, famous for its firm Cabernet Sauvignon, sits a bit higher: the name is trademarked by Bodega Norton. The fruit leans toward deep strawberry, with notes of minerality and clove. Las Compuertas is an official IG and one of Luján de Cuyo's finest districts, with altitudes approaching 3,600 feet and stony soils mixed with abundant clay. Ripe red fruit, mostly strawberry and raspberry, typifies the red wines. Another official IG, Agrelo, produces some of the region's best fruit. Some spots that are higher in elevation are often referred to as Alto Agrelo. Wines from Agrelo are marked by freshness, deep black-cherry fruit, and lifted aromatics, which are particularly punctuated in the Malbec's violet nose. The Vistalba district, currently trademarked by Bodega Pulenta, has the stoniest soils. Its vineyards extend up to 3,440 feet. Vistalba wines tend to be tighter and need time in bottle.

RECOMMENDED PRODUCERS

Achaval Ferrer

Alta Vista

Altos Las Hormigas

Belasco de Baquedano

Benegas

Bodega Aleanna

Bodega Norton

Bodega Ruca Malén

Bodega Septima

Bodega Vistalba

Bodegas Escorihuela Gascón

Bodegas Esmeralda

Bodega y Cavas de Weinert

Bressia

Cabernet de los Andes (*see also* Luigi Bosca—Familia Arizu)

Catena Zapata

Chakana

Chandon Argentina

Cheval des Andes

Cruzat

Decero

Dominio del Plata

Doña Paula

Durigutti

Enrique Foster

Ernesto Catena Vineyards

Fabre Montmayou

Goulart

Kaiken

Krontiras

Lagarde

Luca

Luigi Bosca—Familia Arizu

Melipal

Mendel

Nieto Senetiner

Piattelli Vineyards

Poesía

Pulenta Estate

Renacer

Terrazas de los Andes

Trapezio

Viña Alicia (*see also* Luigi Bosca—Familia Arizu)

Viña Cobos

Maipú

Established in 1858 and named for the battle of Maipú (in Chile) during the South American wars against the Spanish, this region produces red wines that are typically less powerful than those of Luján de Cuyo while still maintaining excellent fruit and a fleshy, elegant mid-palate. About the same size as Australia's Barossa Valley, this department, which is also an IG, has an area of 34,348 acres. The plantings are 24 percent Malbec, 13 percent Cabernet Sauvignon, 7 percent Merlot, and 7 percent Pedro Giménez, among other grapes. Of the two main districts, Lunlunta is better known. Containing many excellent parcels, its vineyards are at about 2,800 feet and have sandy soils. Its wines are ripe in character, with sweeter tannins and flavors akin to ripe raspberry and Bing-cherry yogurt. Lesser-known Barrancas, also an IG of Maipú, has soils that are even sandier, and it produces wines with a firm, ripe red-fruit character. The remaining regions, Russell and El Paraíso, produce wines much like Lunlunta's in style.

RECOMMENDED PRODUCERS

Andean Viñas (*see* Grupo Peñaflor)

Antigal

Familia Cecchin

Familia Zuccardi

Finca Flichman

López

Manos Negras

Michel Torino (*see also* Grupo Peñaflor)

Pascual Toso

Ricardo Santos

Rutini

Trapiche (*see also* Grupo Peñaflor)

Trivento

Uco Valley

A *zona interjurisdiccional* of three departments, the Uco Valley, a bit more than an hour by car southwest of Mendoza city, is the "It girl" of the moment. With slopes surpassing five thousand feet, the Uco Valley is spread among several districts in the Graven floodplain. From the Tupungato department in the north to San Carlos in the south, the Uco Valley is roughly forty-five miles long and an average of fifteen miles wide, along the northerly course of the Tunuyán River. The municipality of Tunuyán, whose population is around 45,000, is in the heart of the region, on the western bank of the eponymous river. At 55,400 acres, the Uco Valley is slightly smaller than the planted area of California's Sonoma County. Over 33 percent of the grapes here are Malbec, and most of the region is newly planted, with drip irrigation. In this location near the Andes, frost is the big concern, and veterans of the region claim that about every five years, a hard spring can cause the loss of up to 30 percent of the vintage.

As an epicenter of viticultural experimentation, the Uco Valley has hosted many trials seeking the best matches of soils with clones and rootstocks. The red wines, which account for 80 percent of the cultivation, feature power, minerality, linear persistence, gritty tannins, and sharp acidity. The higher altitudes produce fruit with deeper pigmentation and thicker skins.

Tupungato   At the northern end of the Uco Valley, the department of Tupungato is located at 33° south. Established in 1858, the department derives its name from the nearby Tupungato volcano, one of Argentina's highest peaks at 21,555 feet, and is both an administrative department and an IG. Tupungato, meaning "star viewpoint" in the native Huarpe language, is broken down into several districts. San José, an area whose vineyards extend as high as 4,260 feet, has shallow, sandy loam soils with some stones. It makes very plummy and blackberry-scented wines. Neighboring Villa Bastías is similar in profile, with wines that have black fruit and tea-leaf flavors. With approachable tannins, these grapes are used more for blending than for IG-labeled wines. Nearby El Parral's wines are like those of San José, with spice and licorice notes.

With vineyards at heights of more than five thousand feet, Gualtallary is the star district of the region. Long and narrow with very stony, limestone-flecked soils, the northern portion is said to be best, and the red wines are characterized by smoky blackberry and damson plum flavors and distinct chalky minerality. The limestone also makes Gualtallary the source of the region's most distinctive whites. Rounding out the region is Cordón del Plata, a district that produces wines similar in style to those of Villa Bastías. Ruta 89, the road where almost all the well-known producers are located, is the local equivalent of the Napa Valley's Silverado Trail.

San Carlos   Established in 1772, San Carlos is located on the southeastern edge of the Uco Valley. The terrain is largely alluvial and ideal for easily-managed viticulture. With around the same planted area as Australia's McLaren Vale, the region is divided into four districts. Home to most of the Uco's older vines, La Consulta has heavier soils, and its primarily flood-irrigated vineyards are known for producing dense black fruit with attractive herbal notes and excellent structure. Altamira, sitting at close to 3,600 feet with sandy, stony soils, is a great source of savory wines reminiscent of those from Tupungato's Gualtallary. Punctuated with black-fig, peppercorn, and marmalade notes, the wines from Altamira are powerful and age-worthy. The wines of the Eugenio Bustos district are much like those of La Consulta but with less complexity. The lesser-known but excellent El Cepillo can produce first-rate fruit resulting in floral wines, scented with raspberry and blueberry fruit, but it is a cold area, more susceptible to frost than the rest of Mendoza, and grapes ripen late.

Tunuyán The department of Tunuyán was established in 1880. It forms the third portion of the Uco Valley, with 18,000 planted acres spread among four districts. Los Chacayes is administratively part of Vista Flores and currently a registered trademark of Bodega Lurton. With minimal topsoil, these rocky vineyards produce wines that are very floral (with essence of violets) and ooze fruit. Washed stones fleck the vineyards, and the wines have sour-cherry notes and are less aromatic than those of neighboring vineyards. The area is a source of some of Mendoza's best Merlot, a variety often used as a blending grape in Argentina. Sitting between three and four thousand feet above sea level, Vista Flores, Tunuyán's most celebrated commune, makes similarly styled but usually riper and more complex wines than its administrative neighbor. Los Árboles and Los Sauces are less distinctive subareas.

RECOMMENDED PRODUCERS

Altocedro	Flechas de los Andes
Andeluna	La Posta
Antucura	Masi Tupungato
Bodega Atamisque	Monteviejo
Bodega Piedra Negra	Mumm Argentina
Bodegas Caro	O. Fournier Argentina
Callia	Passionate Wine
Clos de los Siete	Riglos
Cuvelier Los Andes	Rolland Collection
DiamAndes	Salentein
Finca La Celia	Vistaflores
Finca Sophenia	Zorzal

San Rafael

San Rafael is the anchor of the Southern Mendoza region. Situated 125 miles to the south of Mendoza city, the city of San Rafael is the second largest in Mendoza and has an oasis wine district that sits between the Diamante and Atuel rivers in an otherwise desert terrain. In San Rafael, which has one of two of Mendoza's DOC designations (the other is Luján de Cuyo), cultivable land is very limited because of the lack of water. Because the region is also vulnerable to hail, preventive netting is used on the vines. More renowned for ecotourism than for wine, this region sees close to 100,000 tourists annually, most of them from other parts of the country, and sells 80 percent of its wines within Argentina. At a lower altitude and with a built-in market, San Rafael has eschewed much of the internationally focused marketing of its Mendoza neighbors. San Rafael is warmer than much of Mendoza and the fruit has less intensity, but the wines are ripe, easy-drinking, tasty, and approachable. There are close to twenty wineries here, but only a few are considered serious players.

RECOMMENDED PRODUCERS

Alfredo Roca
Casa Bianchi
Finca Dinamia

PATAGONIA

A thousand miles southeast of Buenos Aires, Patagonia derives its name from the word *Patagón,* used by Ferdinand Magellan to describe the native people, whom he thought to be giants (the native Tehuelches were nearly six feet tall, compared to the Spaniards at just over five feet). The territory, a *región interjurisdiccional,* is made up of four provinces—La Pampa, Neuquén, Río Negro, and Chubut. The official administrative region also encompasses Santa Cruz and Tierra del Fuego farther south. Although 30 percent of Argentina's land is in greater Patagonia, only 5 percent of the population lives in the region. With multiple ski resorts, fabulous fly fishing, and ice fields that constitute one of the world's largest reserves of freshwater, Patagonia is a land of immense plains, deserts, mountains, and breathtaking vistas. With perennially strong breezes and Antarctic winds, which have led to the widespread planting of poplar trees as barricades, there are no insect pests. The latitude (rather than the altitude) creates conditions of intense sunlight, low rainfall, and wide thermal amplitude. With pristine wine conditions, Patagonia has bright prospects for growing grapes (as well as other fruit—you've never tasted a red apple until you've sampled one from Río Negro). Patagonia is distinguished for its citrusy Chardonnay, graphite-scented Malbec, well-defined Cabernet Franc, fresh and vibrant Merlot, and the best Pinot Noir in the country. On the whole, Río Negro reds tend to be weightier and spicier than their Neuquén brethren.

Winemaking in Patagonia is concentrated in Río Negro and Neuquén, directly to the west. At 850 feet above sea level and with 80 percent of the vineyards, Río Negro's

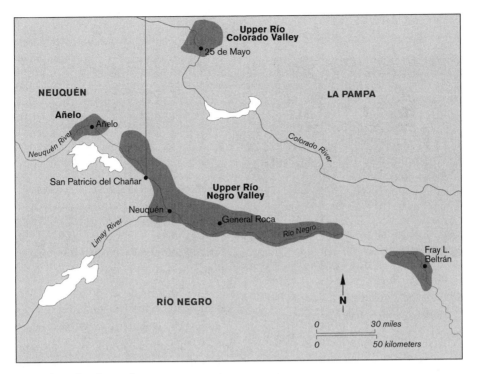

Map 5. Argentina: Patagonia

upper valley (Valle Alto) has four wine IGs: Avellaneda, General Conesa, Pichimahuida, and the best-known of the quartet, General Roca. Some of the vines of Río Negro date back over one hundred years. At nine hundred feet above sea level and approximately 38° south, Neuquén has 4,500 acres planted, about 50 percent less than Río Negro. On the border with Chile, Neuquén has five regions: of these, Chañar-Añelo, near the town of San Patricio del Chañar, is the most important.

At this point, the La Pampa province is home to very little winemaking: it focuses on the celebrated cattle farming of the Argentinean plains. Its one wine region, of less than one hundred acres, is located in 25 de Mayo. Chubut, too, is very small at this point but very promising, especially with Riesling. Quality winemaking in Patagonia is concentrated in Río Negro and Neuquén.

RECOMMENDED PRODUCERS

Bodega del Fin del Mundo

Bodega Noemía

Bodega Patritti

Chacra

Desierto

Familia Schroeder

Humberto Canale

Marcelo Miras

Patagonian Wines

WINERY PROFILES

XumeK's vineyards in the Zonda Valley, San Juan, Argentina. Photo by Laura Mariategui.

Achaval Ferrer

Founded: 1998

Address: Calle Cobos 2601, Perdriel, Mendoza, www.achaval-ferrer.com

Owners: Santiago Achaval Becú, Manuel Ferrer Minetti, Marcelo Victoria, Diego Rosso, Tiziano Siviero, Roberto Cipresso

Winemaker: Robert Cipresso

Viticulturist: Diego Rosso

Known for: Malbec

Signature wine: Finca Altamira

Tasting: Open to the public by appointment only; tasting room

When you meet the accomplished Santiago Achaval, you know in a moment why he and his project are considered among the best in all of South America. The winery's accomplishments, too many to list, are aptly summarized by the *Wall Street Journal,* which declared Achaval Ferrer to be "arguably Argentina's first 'cult' winery" and Santiago as "one of Argentina's most revered winemakers." Working from seven vineyards, the impressive team includes Argentine partner Manuel Ferrer and the Italian dynamic duo of Roberto Cipresso and Tiziano Siviero, who own La

Fiorita Winery in Montalcino. The quartet of single-vineyard wines (Altamira, Bella Vista, Mirador, and Quimera) are sensational. Cipresso has two fun projects in his Saltena "Altesa" and Italian "Wine Circus," which, though not well known, are well worth checking out, as are Achaval's "Hand of God" wines. The winery restaurant and the tasting room in Chacras merit a visit.

Alfredo Roca

Founded: 1976

Address: Ruta Provincial 165, Cañada Seca, San Rafael, Mendoza, rocawines.com

Owner: Roca family

Winemaker: Raúl Arroyo, Alejandro Roca

Viticulturist: Alfredo Roca

Known for: Malbec, Pinot Noir

Signature wine: Fincas Pinot Noir

Other labels: Preciado, Roca, Family Reserve

Visiting: Open to the public; tasting room

A fourth-generation winery, Alfredo Roca works 280 acres across four vineyards owned by the Roca family: La Perseverancia, Los Amigos, Las Paredes, and Santa Herminia. It makes a range of wines, including some varietal wines you don't often find: Tocai Friulano and Chenin Blanc.

Alta Vista

Founded: 1998
Address: Alzaga 3972, Chacras de
    Coria, Luján de Cuyo, Mendoza,
    www.altavistawines.com
Owner: D'Aulan family
Winemaker: Philippe Rolet
Viticulturist: Pamela Alfonso
Known for: Malbec, Torrontés
Signature wine: Alto
Other labels: Terroir Selection,
    Atemporal
Visiting: Open to the public; tasting room;
    restaurant

Many people refer to Alta Vista as the one that got away, a reference to its withdrawal from the Clos de los Siete project established by Michel Rolland. They did not renege on a commitment; they simply felt that they already had a working winery and decided to focus on the vineyard and on obtaining quality fruit. With 545 acres across five Mendocino vineyards and one in Cafayate, the winery produces about 65 percent Malbec and Malbec-related wine. They are beloved both for their wines and for their agreement to share their "single vineyard" trademark with the entire industry. Alta Vista's single-vineyard wines are iconic.

Altocedro

Founded: 2001
Address: Ejército de los Andes at Tregea 1,
La Consulta, San Carlos, Mendoza,
    www.altocedro.com.ar
Owner: Karim Mussi Saffie
Winemaker: Karim Mussi Saffie
Viticulturist: Guillermo Cacciaguerra
Known for: Malbec, Tempranillo, Cabernet
    Sauvignon
Signature wine: Altocedro Gran Reserva
    Malbec
Other labels: Año Cero, La Consulta Select
Tasting: Open to the public by appointment
    only

Located in La Consulta at over 3,600 feet above sea level, Altocedro has three different vineyards covering fifty-five acres, all of which are between fifty and seventy years old. Altocedro means "tall cedar" in Spanish, and the winery was named for the massive cedar trees that tower over it. Practicing what it calls " noninvasive" farming, this winery has become a cultish favorite in a short time. Karim Saffie comes from a wine family and is considered a young star in Argentina. In addition to working with his own winery, Karim consults for others, and in 2007, the Consejo Empresario Mendocino recognized him with the award for Outstanding Young Mendocino.

Altos Las Hormigas

Founded: 1995
Address: La Legua sin número, Colonia
    Los Amigos, Medrano, Luján de Cuyo,
    Mendoza, www.altoslashormigas.com
Owners: Alberto Antonini, Antonio
    Morescalchi
Winemakers: Alberto Antonini, Attilio
    Pagli
Viticulturist: Mauricio González

Viticulture consultant: Pedro Parra

Known for: Malbec

Signature wines: Malbec Reserva, Valle de Uco

Other labels: Colonia Las Liebres

Visiting: Open to the public by appointment only

The property was named for the ants (*hormigas*) whose colonies were real obstacles to cultivation. Open fields with countless nests surround the estate, and the ants loved to feed on the tender sprouts of newly planted vines. Refusing to poison the ants, the owners looked for natural ways to deter them and commented that the ants were the real owners of the place. As the vines grew, the ants moved on to other foods. The Chilean terroir sleuth Pedro Parra advises the globally respected team of Alberto Antonini and Attilio Pagli. Their wildly successful sister brand, Colonia Las Liebres, is a benchmark for Bonarda.

Amalaya

Founded: 2010

Address: 25 de Mayo sin número, Cafayate, Salta, www.amalaya.com

Owner: Hess family estate

Winemaker: Francisco Puga

Consulting winemaker: Randle Johnson (Hess family winemaker)

Viticulturist: Javier Grané

Known for: Malbec, Torrontés

Signature wine: Amalaya Gran Corte

Visiting: Open to the public by appointment only; tasting room

Amalaya is a new winery project owned by Donald Hess and formerly made at Co-

lomé (see below) before Hess bought Bodega Familia Muñoz and its operations. Amalaya means "hope for a miracle" in the language of the Calchaquí Valley indigenous people. Located in the region of Divisadero in the Upper Cafayate, the winery makes only blends (*cortes*).

Andeluna

Founded: 2003

Address: Ruta Provincial 89, km 11, Gualtallary, Tupungato, Mendoza, www.andeluna.com

Owner: Lay family

Winemaker: Manuel González

Consulting winemaker: Michel Rolland

Viticulturist: Manuel González

Known for: Malbec

Signature wine: Pasionado

Other labels: Asionado, Altitud, 1300

Visiting: Open to the public by appointment only; tasting room; restaurant

Andeluna Cellars was named for the romance and heritage of the Argentine moon illuminating the Andes Mountains above the Uco Valley near the town of Tupungato. It's owned by the well-known international businessman H. Ward Lay (son of the late Herman W. Lay, founder of Frito-Lay and former chairman of the board of PepsiCo). Lay has many interests, especially in Argentina. The winery receives over five thousand visitors annually and offers cooking classes.

Antigal Winery & Estates

Founded: 1999

Address: Calle Maza at Calle Manuel A.

Sáenz, Russell, Maipú, Mendoza,
www.antigal.com

Owner: Peiro family

Winemaker: Miriam Gómez

Viticulturist: Cristina Herrera

Known for: Malbec

Signature wine: Antigal ONE "La Dolores"
Vineyard

Other labels: Aduentus, Cavia

Visiting: Open to the public by appointment only; tasting room

In the shell of a winery that dates back to 1897, the Peiros established Antigal in an architectural style that pays homage to tradition. Inside, the winery is state of the art, with equipment from France and Italy. The use of a gravity-flow system has eliminated the need for pumps and other forceful handling. As is often the case in Mendoza, winery and vineyard locales are varied: there are three vineyards in Tupungato in the Uco Valley and another four at and around the winery in Maipú, only sixteen miles from the city of Mendoza.

Antucura

Founded: 1993

Address: Calle Barrandica sin número,
Vista Flores, Tunuyán, Mendoza,
www.antucura.com

Owner: Caviar Bleu SA (Anne-Caroline
Biancheri)

Winemaker: Hervé Chagneau

Consulting winemaker: Michel Rolland

Viticulturist: Matias Cazorla

Known for: Malbec, Cabernet Sauvignon,
Merlot, Pinot Noir

Signature wine: Antucura Grand Vin

Other labels: Barrandica

Visiting: Restaurant; hotel

Bodega Antucura is owned by the French-born publisher Anne-Caroline Biancheri. Her 100-acre vineyard is located in the prestigious Vista Flores region in Tunuyán in the Uco Valley, an area with ideal climatic and geographic conditions for the development of flagship Argentinean wines. The word *Antucura*, which means "sun stone" in the Mapuche language, refers to the glacial stones that line the vineyard. Spending a night at Casa Antucura is like visiting your wealthiest friends at their country home. The eight-suite inn supplies privacy, highbrow art and books, superb wine, and enchanting Andean vistas.

Belasco de Baquedano

Founded: 1910 (vineyards); 2008 (current
winery)

Address: Calle Cobos 8260, Luján de Cuyo,
Mendoza, www.belascomalbec.com

Owner: Juan Ignacio Belasco

Winemaker: José Ponce

Consulting winemaker: Bertrand Bourdil

Viticulturist: Belasco de Baquedano Team

Known for: Exclusively Malbec

Signature wine: Swinto Old Vine Malbec

Other labels: Swinto, AR Guentota,
LLAMA, Rosa

Visiting: Open to the public; tasting room;
restaurant (Navarra)

The Spaniard Juan Ignacio Belasco, president of Grupo La Navarra, is scion to the Belasco family, which also owns four wineries in Spain. Focusing exclusively on Malbec, the winery, whose name comes from the Huarpe words meaning "crow" and "Cuyo's soul," is off the beaten track but well worth the journey. Inside the state-of-the-art winery you'll find a long hallway known as the Aroma Room. Along the walls are forty-eight stations, each emitting its own distinctive aroma from an oil capsule when you flip a lever. Belasco lies in Mendoza's finest vineyard zone, Agrelo, a subregion of Luján de Cuyo. The winery is surrounded by 170 acres of hundred-year-old vines in Agrelo and Perdriel. The restaurant overlooks the vineyards.

Benegas

Founded: 2000

Address: Aráoz 1600, Mayor Drummond, Luján de Cuyo, Mendoza, www.bodega benegas.com

Owner: Federico Benegas Lynch

Winemaker: Federico Benegas Lynch

Viticulturist: Darío Burgos

Known for: Malbec, Cabernet Franc, Cabernet Sauvignon

Signature wine: Lynch Meritage

Visiting: Open to the public by appointment only

In the late 1800s, Tiburcio Benegas was a pioneer in Argentinean wine, responsible for the first significant plantings and cataloging of French grapevines and promoting the most advanced technology available at the time. Benegas Lynch, a former board member of Grupo Peñaflor, is a direct descendant of Benegas who returned to his roots in wine production in 1998. The original adobe winery was built in 1901 and was once owned by a governor of Mendoza. Their vineyard, Finca Liber-tad, is situated in the southeast corner of Cruz de Piedra and is part of the original Benegas family property, with some vines more than one hundred years old.

Bodega Aleanna

Founded: 2007

Address: Ruta Provincial 92 sin número, Luján de Cuyo, Mendoza, www.elenemigowines.com

Owners: Adrianna Catena, Alejandro Vigil

Winemaker: Alejandro Vigil

Viticulturist: Alejandro Vigil

Known for: Malbec, Cabernet Franc, Petit Verdot, Bonarda, Chardonnay

Signature wines: El Enemigo Malbec, El Gran Enemigo

Visiting: Open to the public by appointment only; tasting room

These cult wines are the results of a collaboration between Nicolás Catena's youngest daughter, Adrianna, and the chief winemaker, Alejandro Vigil. Besides being a respected winemaker, Alejandro Vigil knows his terroir; he was the former head of the soil division at Argentina's Wine Institute. The winery currently makes two notable red wines: El Enemigo, a blend of Malbec and Petit Verdot, and El Gran Enemigo, which is based on Cabernet Franc. El Enemigo means "the enemy": here it evokes the aphorism that the most important battle is the one fought with the original enemy, yourself.

Bodega Atamisque

Founded: 2006

Address: Ruta Provincial 86, San José,

Tupungato, Mendoza, www.atamisque
.com

Owner: Jean du Monceau

Winemaker: Philippe Caraguel

Viticulturist: Philippe Caraguel

Known for: Malbec, Cabernet Sauvignon,
Pinot Noir, Merlot, Chardonnay

Signature wine: Atamisque Malbec

Other labels: Catalpa, Serbal, Cave
Extreme, Paul Rigaud

Visiting: Open to the public by appointment only; tasting room; restaurant
(Rincón Atamisque); lodge

Owned by a Frenchman, Jean du Monceau, who purchased the former Estancia
Atamisque in 2006, the winery is stunning, looking like a French or Italian al-
pine house. With its 1,730 acres of vineyards, fruit orchards, park, and open land,
it is an oasis in the middle of Argentina.
The property features a nine-hole golf
course with trees planted seventy years
ago. A new guest lodge, sited among the
vines, is supposed to be spectacular. Atamisque is popular for its very good restaurant and beautiful landscape, which includes a trout pond providing fish for the
daily menu.

Bodega Augusto Pulenta

Founded: 2000

Address: General Acha 982 Sur, San Juan,
www.augustopulenta.com

Owners: Mario Augusto Pulenta, Mario
Daniel Pulenta, María Andrea Pulenta,
and María Gabriela Pulenta

Winemaker: Hugo Angel Torres

Consulting winemaker: Paul Hobbs

Known for: Cabernet Sauvignon, Malbec,
Bonarda, Syrah

Signature wine: Augusto P. Blend

Other labels: Valbonda Tradición

Visiting: Open to the public by appointment only

The Pulenta family is one of the most far
reaching in the country, with three distinctive operations run by descendants of
Angelo Polenta, who changed his name to
Pulenta when he arrived in San Juan from
Buenos Aires in 1902. He bought a plot
of land that, in 1910, became the world's
largest winery and ultimately the anchor
of Grupo Peñaflor, the parent company
of Trapiche and now one of the world's
largest wine and beverage companies. In
1997, his grandson Mario decided to sell
his stock in the parent company and start
again in the heart of the Tulum Valley,
making more traditional styles of wine in
partnership with his three children.

Bodega del Fin del Mundo

Founded: 2000

Address: Ruta Provincial 8, km 9,
San Patricio del Chañar, Neuquén,
www.bodegadelfindelmundo.com

Owners: Julio Viola, Eduardo Eurnekian

Winemaker: Marcelo Miras

Consulting winemaker: Michel Rolland

Viticulturist: Pepe Barria

Known for: Pinot Noir, Malbec, Merlot,
Cabernet Franc

Signature wine: Fin del Mundo Special
Blend Reserva

Other labels: Postales, Newen, Ventus,
Cosecha de Mayo, Fin

Visiting: Open to the public; tasting room; restaurant

Fin del Mundo is responsible for the existence of the Neuquén wine region. A successful real-estate man with a passion for wine, Julio Viola did extensive computer modeling to identify the ideal area for viticulture on Neuquén's arid plateau, about 1,200 feet above sea level. He convinced the government of Argentina to lend him enough money to purchase more than three thousand acres. After studying drip irrigation in Israel, he built a twelve-mile canal to carry water from the Neuquén River to his land, a project that required the installation of several pumping stations and thousands of miles of pipe. In early 2012, Viola sold half of his endeavor to the Argentinean businessman Eduardo Eurnekian, and together they bought NQN winery. Following the success of their Experiencia del Fin del Mundo restaurant in the Palermo neighborhood of Buenos Aires, they plan to open others.

Bodega Noemía

Founded: 2001

Address: Ruta Provincial 7, km 12, Valle Azul, Río Negro, www.bodeganoemía .com

Owners: Noemi Marone Cinzano, Hans Vinding-Diers

Winemaker: Hans Vinding-Diers

Viticulturist: Hans Vinding-Diers

Known for: Malbec

Signature wine: Bodega Noemía Malbec

Other labels: J. Alberto, A Lisa, Noemía "2"

Visiting: Open to the public by appointment only; tasting room

Bodega Noemía de Patagonia is the project of Countess Cinzano and her partner, the South African winemaker Hans Vinding-Diers, who were inspired by the potential of the region after tasting exceptional wines of the area from the 1930s and 1940s. A scouting mission of the Río Negro Valley resulted in the purchase of the semiabandoned Bodega Noemía property and significant restoration of the vineyards. The estate employs fermentation with native yeasts exclusively, and Noemía's wines are considered among Argentina's finest. In addition to working at Noemía, Vinding-Diers is winemaker at Chacra.

Bodega Norton

Founded: 1895

Address: Ruta Provincial 15, km 23.5, Perdriel, Luján de Cuyo, Mendoza, www.norton.com.ar

Owners: Gernot Langes-Swarovski, Michael Halstrick

Winemaker: Jorge Riccitelli

Viticulturist: Pablo Minatelli

Known for: Malbec, Cabernet Sauvignon, Torrontés, Chardonnay

Signature wines: Norton Reserve Malbec, Gernot Langes

Other labels: Norton, Mini, Lo Tengo, Vistaflor, Quorum, Finca Perdriel

Visiting: Open to the public; tasting room; restaurant (La Vid)

The Norton winery has occupied its site for more than a century. It is named for

Edmund Norton, an English engineer who planted vines at the estate in 1895 but is better known for his role in building the railway connecting Mendoza with Chile. Norton is owned today by the Swarovski family, of the Austrian crystal dynasty: its patriarch, Gernot Langes-Swarovski, is a wine fanatic who makes his own wine in the South Tirol. He fell in love with this area of Argentina in the late 1980s and was impressed enough to buy a winery which now works 1,680 acres of vineyard. Known since the early 1980s for its Malbec, Norton produces Finca Perdriel, one of Argentina's iconic reds. Norton was awarded a Best of Wine Tourism award in 2007 by the Great Wine Capitals Association.

Bodega Patritti

Founded: 2003

Address: Picada 1 norte, San Patricio del Chañar, Neuquén, www.bodegaspatritti .com.ar

Owner: Rubén Patritti

Winemakers: Mariano di Paola, Nicolás Navío

Known for: Malbec

Signature wine: Primogénito

Other labels: Lassia

Visiting: Open to the public by appointment only; tasting room

Designed by the prestigious architectural firm of Sidoni & Associates, this winery is both pleasing to the eye and at the cutting edge of technology. The structure forms a great wave shape, evoking the movements of ancient water flows. In 2010 Patritti opened shops in Patagonia and Bue-

nos Aires. In 2011, the winery helped launch a wine-tourism route highlighting Neuquén's wine, apples, and dinosaur fossils. None of these sidelines take attention away from the wines, which are quite good.

Bodega Piedra Negra (François Lurton)

Founded: 1996

Address: Ruta Provincial 94, km 21, 5565 Vista Flores, Tunuyán, Mendoza, www.francoislurton.com

Owner: François Lurton

Winemaker: Thibault Lepoutre

Viticulturist: Julio Chaab

Known for: Malbec, Pinot Gris

Signature wine: Piedra Negra Malbec

Other labels: Chacayes, Gran Lurton, Pasitea

Visiting: Open to the public by appointment only; tasting room

If you're looking for Bodega Lurton, head for Bodega Piedra Negra (François Lurton)! Following the success of its iconic Malbec, Piedra Negra, François Lurton decided in 2012 to change the winery name to something that would express the character of the 130 acres of vineyard in Vista Flores, so he named it for the black stones that characterize the vineyard soils. His father, André Lurton, owns seven châteaux in France and is the patriarch of one of the first families of Bordeaux. François and his brother Jacques have additional winery projects in the Languedoc, Spain, and Chile, each with its own team of winemakers.

Bodega Poesía

Founded: 2000

Address: Castro Barros sin número, Mayor
   Drummond, Luján de Cuyo, Mendoza,
   www.bodegapoesia.com

Owners: Hélène-Garcin Lévêque, Patrice
   Lévêque

Winemakers: Hélène-Garcin Lévêque,
   Patrice Lévêque

Viticulturist: Patrice Lévêque

Known for: Malbec, Cabernet Sauvignon

Signature wine: Poesía

Other labels: Cuvée Hélène, Clos des
   Andes, Pasodoble

Visiting: Open to the public; tasting room

Hélène Garcin-Lévêque and her husband,
the viticulturist Patrice Lévêque, are owners of some of Bordeaux's most sought-
after labels, including Clos l'Église in
Pomerol, Châteaux Branon and Haut-Ber-
gey in the Graves, and Château Barde-
Haut in Saint-Émilion. They have thirty-
two acres of land in Mendoza, with the
majority planted to Cabernet Sauvignon
and Malbec from vines dating back to
1935, and the winery is dedicated to producing high-end, limited-edition wines.
The vineyard is managed organically.

Bodega Ruca Malén

Founded: 1999

Address: Ruta Nacional 7, km 1059,
   Agrelo, Luján de Cuyo, Mendoza,
   www.bodegarucamalen.com

Owners: Jacques Louis de Montalembert,
   Jean Pierre Thibaud

Winemaker: Pablo Cuneo

Viticulturist: Pablo Cuneo

Known for: Malbec, Cabernet Sauvignon

Signature wine: Kinien de Don Raúl

Other labels: Yauquén, Kinien

Visiting: Open to the public; tasting room;
   restaurant

In the Mapuche language, *ruca malén*
means "the house of the young girl." The
name comes from a local myth of the romance between a god and a fearless, beautiful woman. Legend has it that the woman
looked into the eyes of the god and fell in
love, but he could not stay on the earth. So
he built her a house and left her an elixir
by which she could remember him. The
winery, a partnership between Thibaud,
former CEO of Chandon Argentina, and
his Burgundy-born friend de Montalembert, is considered one of the most hospi-
table in Mendoza, and the restaurant alone
is worth the trip.

Bodega Septima

Founded: 2001

Address: Ruta Nacional 7, km 6.5,
   Agrelo, Luján de Cuyo, Mendoza,
   www.bodegaseptima.com

Owners: Grupo Codorníu, Raventos family

Winemaker: Paula Borgo

Consulting winemaker: Arthur O'Connor

Viticulturist: Alejandro Livellara

Known for: Malbec, Cabernet Sauvignon,
   Syrah, Tempranillo, Tannat,
   Chardonnay, Sauvignon Blanc,
   Gewürztraminer

Signature wine: Septima Gran Reserva

Other labels: María Codorníu

Visiting: Open to the public; tasting room; restaurant (María Restaurant)

The 750-acre estate winery is named for its position as seventh winery of the Codorníu group's empire. Septima's dramatic building was designed by the eminent Mendoza architects Eliana Bórmida and Mario Yanzón. It follows ancient Inca construction principles, by which natural Andean stone is cut, shaped, and stacked to make drystone walls. The resulting appearance contrasts with the polished look of so many wineries. Paula Borgo worked at both Kendall-Jackson and Doña Paula before coming to the winery. She is advised by Arthur O'Connor, director of enology for Grupo Codorníu.

Bodega Tacuil

Founded: 1999

Address: Tacuil, Molinos, Salta, www.tacuil.com.ar

Owner: Dávalos family

Winemaker: Raúl Dávalos Jr.

Known for: Malbec

Signature wine: Vinas de Dávalos

Visiting: Open to the public by appointment only

The Dávalos family has crafted Tacuil wines since the early 1800s, with fourth-generation Raúl Dávalos now refining their offerings. This is one of the rare instances of a winery's having its own appellation (Valle Tacuil). The winery uses no oak and ferments only with native wild yeasts. It is literally off the grid (bring your GPS!) at an altitude of nearly 8,500 feet above sea level, miles away from towns, restaurants, and the tourist routes. Plan ahead for the trip: it's some three and a half hours from Cafayate with no rest stops, no convenience stores, and no ice-cream stands.

Bodega Tamarí

Founded: 2005

Address: Circunvalación Celia Bustos de Quiroga 374, Eugenio Bustos, San Carlos, Mendoza, www.tamari.com.ar

Owner: VSPT Wine Group Argentina

Winemaker: Sebastián Ruiz

Viticulturist: Manuel Bianchi

Known for: Malbec, Cabernet Sauvignon, Torrontés

Signature wine: Tamarí Zhik

Visiting: Open to the public by appointment; tasting room; restaurant (Posada La Celia)

Meaning "to do everything with passion" in Huarpe, Tamarí is the second Argentinean winery of Viña San Pedro Tarapacá (VSPT), Chile's second-largest wine company. Tamarí began as an export-focused project and only began selling domestically in 2010. Its wines are perennial good values.

Bodega Vistalba (Carlos Pulenta Wines)

Founded: 2001

Address: Roque Saenz Peña 3531, Vistalba, Mendoza, www.carlospulentawines.com

Owner: Carlos Pulenta

Winemaker: Alberto Antonini

Viticulturist: Facundo Yazli

Known for: Malbec

Signature wine: Vistalba Corte A

Other labels: Tomero, Progenie

Visiting: Open to the public by appointment only; tasting room; inn (La Posada) and restaurant

A member of the extensive Pulenta clan, Carlos, the former president of Salentein Family Wines, is the very successful leader of Vistalba. Located in the heart of the eponymous area, the winery labels its blends (*cortes*) as simply A, B, and C based on their components. The restaurant, La Bourgogne, under the capable care of chef Jean-Paul Bondoux, is considered one of the region's finest. The operation's hospitality overall is outstanding. Having sold half his shares of the winery in 2012, Pulenta is now a partner with the Argentinean billionaire Alejandro Pedro Bulgheroni, who owns Uruguay's Garzón winery, as well as California's Renwood Winery.

Bodegas Caro

Founded: 1999

Address: Presidente Alvear 151, Godoy Cruz, Mendoza, www.bodegascaro.com

Owner: Les Domaines Barons de Rothschild (Lafite), Catena family

Winemaker: Fernando Buscema

Viticulturist: Fernando Buscema

Known for: Cabernet Sauvignon, Malbec

Signature wine: Caro

Other labels: Amancaya, Aruma

Visiting: Open to the public by appointment only

Bodegas Caro is a Catena family partnership with Domaines Barons de Rothschild, the owners of Château Lafite Rothschild.

As in France, the operation focuses on classic red Bordeaux blends. The winery is located next to the historic Escorihuela Gascón winery in Mendoza City. Caro's underground caves, where tastings are held, date back to 1884, when the original brick building was constructed. From the inception of the Argentinean project, Lafite has sent over barriques coopered in France. All grapes for this project are purchased.

Bodegas Escorihuela Gascón

Founded: 1884

Address: Belgrano 898, Godoy Cruz, Mendoza, www.bodegasescorihuela.com

Owners: Nicolás Catena, Benegas family

Winemaker: Gustavo Marín

Viticulturist: Gonzalo Videla

Known for: Malbec, Cabernet Sauvignon

Signature wine: Escorihuela Gascón Malbec

Other labels: Don Miguel Escorihuela Gascón, Don Miguel Gascón, Escorihuela Gascón, Familia Gascón, Pequeñas Producciones, the President's Blend, 1884 Reserve, Mocayar, Chukker 1, High Altitude, Circus, Sol Amante, Candela Wines

Visiting: Open to the public; tasting room; restaurant (1884)

Miguel Escorihuela Gascón emigrated from Spain in 1880. He arrived in Buenos Aires and first began to appreciate wine while working in a grocery store. Soon after, he moved to Mendoza and started his own grocery store. From this he earned enough to buy forty-two acres of land, plant vineyards, and start building Bodegas Escorihuela. The successful Gascón finalized the construction of Mendoza's tallest building, Pasaje San Martín (still standing after eighty years), and donated land to build the first wing of the Hospital Español, still one of Mendoza's leading hospitals. The winery is now a partnership between Nicolás Catena and the Benegas family, with Ernesto Catena as president. It houses chef Francis Mallmann's world-famous 1884 restaurant.

Bodegas Esmeralda

Founded: 1959 (Tilia Brand created in 2006)
Address: Calle Martínez sin número, Junín, Mendoza, www.tiliawines.com.ar
Owner: Catena family
Winemaker: Alejandro Viggiani
Viticulturist: Alejandro Viggiani
Known for: Malbec
Signature wine: Tilia Malbec
Visiting: Open to the public by appointment only

Tilia wines are made at Bodegas Esmeralda, the Catena family's winery in the east of Mendoza, and are blends of fruit from the Uco Valley and the east of Mendoza. *Tilia* is the Latin name for the linden tree, the source of leaves and flowers for a calming herbal tea popular among Mendocinos. Tilia is dedicated to responsible treatment of the environment. The winemaking and viticulture teams employ sustainable practices and work to promote sustainability in the community. Furthermore, the entire Tilia viticulture team attends regular sustainability training sessions at the National University of Cuyo

and the National Agricultural Research Institute.

Bodegas Etchart

Founded: 1850
Address: Ruta Nacional 40, km 4338, Cafayate, Salta, www.bodegasetchart.com
Owner: Pernod Ricard Argentina
Winemaker: Victor Marcantoni
Viticulturist: Mariano Bustos
Known for: Torrontés, Malbec
Signature wine: Arnaldo B
Visiting: Open to the public; tasting room

The property was established in La Florida in a small country house that still sits in the vineyard. In 1938, Arnaldo Etchart acquired the 161-acre estate. Pernod Ricard bought the winery in 1996 and has been steadily investing in this historic property. Bringing on Michel Rolland as a consultant in 1987 was not only pioneering for the winery but also revolutionary in introducing global standards and palates to the entire Argentinean wine industry.

Bodegas La Guarda

Founded: 2002
Address: Abraham Tapia 1380 Sur, San Juan, www.bodegaslaguarda.com
Owner: Hilda Estela Ciácera
Winemaker: Fabricio Ferrandiz
Viticulturist: Mauricio Frau
Known for: Syrah and Malbec
Signature wine: El Guardado
Other labels: Sangre de Viña, Vika, La Noche
Visiting: Open to the public; tasting room

A third-generation family operation, Bodegas La Guarda is based in the town of Sarmiento, although the winery owns vineyards throughout the region, with a focus on Tulum and Pedernal. Its iconic wine, El Guardado, is a blend of the best of its Syrah and Malbec.

Bodegas López

Founded: 1898

Address: Ozamis 375, General Gutiérrez, Maipú, Mendoza, www.bodegaslopez.com.ar

Owner: López family

Winemakers: Carlos López, Omar Panella

Consulting winemaker: Carmelo Panella

Viticulturist: Diego Cantu

Known for: Malbec, Cabernet Sauvignon, Merlot, Sangiovese, Chardonnay, Sauvignon Blanc, Pinot Noir, Chenin Blanc

Signature wine: Montchenot

Other labels: Federico López, Casona López, Château Vieux, Rincón Famoso, Miguel Brasó, Traful, Vasco Viejo, Xero

Visiting: Open to the public; tasting room; restaurant (Rincón de López); wine museum

Bodegas López is Argentina's largest winery, although it's virtually unknown outside the country. While this position might be expected to provoke jealousy and disdain among other winemakers, the industry generally admires this winery and its historic role. Not long ago, when Argentinean wineries were fewer, larger, and less engaged, producing wine for sale in demijohns and bulk, López was working

hard to deliver consistency and quality, as demonstrated in its Château Montchenot label. However, its style remains proudly classic, with wines that display a highly consistent and distinctive soft palate. The winery incurred huge financial expenses to maintain large volumes of aging wines, readying them for blending and allowing López to offer wines that effectively do not change over time. Argentineans of all socioeconomic strata recognize and love tradition.

Bodega y Cavas de Weinert

Founded: 1975

Address: San Martín 5923, Carrodilla, Luján de Cuyo, Mendoza, www.bodegaweinert.com

Owner: Bernardo Carlos Weinert

Winemaker: Hubert Weber

Known for: Malbec, Cabernet Sauvignon, Merlot, Sauvignon Blanc

Signature wine: Cavas de Weinert

Other labels: Patagonian Wines, Pedro del Castillo

Visiting: Open to the public; tasting room; restaurant by reservation only

Weinert was founded in 1890 by a Spanish immigrant family named Otera, who ran it until 1920. The winery was abandoned as wine consumption fell. In the mid-1970s, Brazilian-born Bernardo Weinert purchased the estate and refurbished it in time for the first harvest of 1977. According to Roberto de la Mota and others, that 1977 Malbec is still a shining example of the grape and one of the most legendary wines ever made in Argentina. At this old-school winery, Swiss winemaker Hubert Weber has access to some of Mendoza's oldest vineyards and has maintained a more traditional style. He makes wines in his spare time for his own eponymous label (see www.hubertweber.com.ar). The Patagonian Wines label is the product of a fifty-acre property in El Hoyo Valley, in Patagonia's Chubut province.

Bressia

Founded: 2001

Address: Cochabamba 7725, Agrelo, Luján de Cuyo, Mendoza, www.bressiabodega .com

Owners: Walter Bressia and family

Winemaker: Walter Bressia

Viticulturist: Walter Bressia Jr.

Known for: Malbec

Signature wine: Profundo

Other labels: Conjuro, Monteagrelo, Lágrima Canela, Última Hoja

Visiting: Open to the public by appointment only; tasting room

Walter Bressia has been in the Argentinean wine trade for almost thirty years. He was the winemaker at Bodega Vistalba and Viniterra from 1998 to 2003 before going out on his own. Oddly, Bressia, one of the best Argentinean winemakers, owns no vineyards. The family works with six growers spread across the Uco Valley and Luján de Cuyo: the winery works the vineyards and buys the grapes but does not own the land. Its white "Lágrima Canela" (Cinnamon Tear) is a fascinating blend of 70 percent Chardonnay and 30 percent Semillon, aged for fourteen months in new French and American oak. Bres-

sia also makes small amounts of delicious grappa and some *méthode traditionnelle* sparkling wine.

Cabernet de Los Andes

Founded: 1901

Address: Mate de Luna 408, 4700 San Fernando dell Valle de Catamarca, Catamarca, www.tizac-vicien.com

Owners: Pedro Vicien, Carlos Arizu

Known for: Bonarda, Malbec, Syrah

Signature wines: Tizac, Vicien

Visiting: Open to the public by appointment only

Cabernet de los Andes is a partnership between Carlos Arizu and Pedro Vicien in the Fiambalá Valley. It was undertaken with a mixture of new plantings and old, scrubby plots into which they have breathed new life. The winery uses only certified organic grapes, farms some of its plots biodynamically, and hopes to go completely biodynamic in time. The strength of the high-altitude sun and lack of water in the porous soils make viticulture challenging. The wines demonstrate Catamarca's potential and are a source of regional excitement.

Callia

Founded: 2003

Address: Avenida de los Ríos at San Lorenzo, Caucete, San Juan, www.bodegascallia.com

Owner: M. P. Wines (Mijndert Pon)

Winemaker: José Rubén Morales

Consulting winemakers: Gustavo Daroni, Ariel Cavalier

Viticulturist: Gustavo Matocq

Known for: Shiraz

Signature wines: Callia Alta, Grand Callia

Other labels: Alta, Amable, Magna

Visiting: Open to the public by appointment only; tasting room

Part of the Mendoza-based Salentein family of wines, Callia is situated between the Pie de Palo hills to the north and the Cerro Chico del Zonda to the south. The winery is named for a young woman who arrived in the land of San Juan many years ago. She wrote to her family describing San Juan as a nirvana replete with abundant fruit despite the desert landscape. Her name was Callia, and today, for the people of Argentina, *Callia* is synonymous with hope, endurance, and success. The attractive Inca-inspired design of the winery is the work of the noted Peruvian stone artist Percy Cuellar.

Casa Bianchi

Founded: 1928

Address: Ruta 143 and Valentín
   Bianchi Street, Las Paredes, Mendoza,
   www.casabianchi.com.ar

Owner: Valentín Bianchi family

Winemaker: Francisco Martinez

Consulting winemaker: Robert Pepi

Viticulturist: Marcelo Garretón

Known for: Malbec, Cabernet Sauvignon,
   Torrontés, Chardonnay

Signature wine: Enzo Bianchi

Other labels: Valentín Bianchi, María
   Carmen, Los Stradivarius de Bianchi,
   Famiglia Bianchi, Leo, Bianchi Particu-
   lar, Génesis, DOC Bianchi, New Age,
   Don Valentín Lacrado, Elsa, Sensual

Visiting: Open to the public; tasting room

Undeniably the San Rafael region's largest and most visible winery, and an ambassador for the industry, Casa Bianchi produces everything from common wines (Bourgogne, Chablis, and Margaux) to some very nice sparkling wines to a full range of premium and ultra-premium wines. More than eighty years ago, when the Italian immigrant Valentín Bianchi bought a vineyard and small winery called El Chiche, could he have dreamed that his grandsons, Valentín Eduardo, Ricardo Stradella, and Alejandro Rubén, would manage one of Argentina's most important wine estates? All Bianchi's wines emanate from three estate properties (totaling 716 acres) in San Rafael. The American Robert Pepi joined the operation as a consultant in 1996. Bianchi sells close to 80 percent of its wines domestically and farms a large area of its vineyards organically, making it one of the larger organic holdings in the country.

Catena Family of Wines (Nicolás Catena)

See Bodegas Escorihuela Gascón, Bodegas Esmeralda, Catena Zapata, Ernesto Catena Vineyards, Tikal

Catena Zapata

Founded: 1902

Address: Calle Cobos sin número,
   Agrelo, Luján de Cuyo, Mendoza,
   www.catenazapata.com, www.alamos
   wines.com

Owner: Catena family

Winemaker: Alejandro Vigil

Viticulturist: Luis Reginato

Known for: Malbec, Cabernet Sauvignon, Chardonnay

Signature wine: Nicolás Catena Zapata

Visiting: Open to the public by appointment only; tasting room

The heart of Nicolás Catena's wine empire is the breathtaking pyramid-shaped Catena Zapata winery. It sources grapes mainly from six historic vineyards, mostly named after family members: Angélica in Lunlunta, La Piramide in Agrelo, Domingo in Tupungato, Adrianna in Gualtallary, Nicasia in Altamira, and Angélica Sur in San Carlos. Catena's winery has been the starting point for many of Argentina's most talented winemakers: Susana Balbo, Pedro Marchevsky, and Alejandro Sejanovich all trained with Catena. Nicolás's amazing daughter Laura is managing director—and a pediatric surgeon. Alejandro Vigil and Luis Reginato, among the best in Argentina, round out the team.

Catena Zapata makes a dizzying array of great wines. In addition, since 1995, the Alamos wines have been among the best values nationally. They are made at the family's winery in Vista Flores with purchased grapes and fruit from younger plantings. They are probably Argentina's most widely sold premium wines.

Chacra

Founded: 2004

Address: Avenida Roca 1945, General Roca, Río Negro, www.bodegachacra.com

Owner: Piero Incisa della Rocchetta

Winemaker: Hans Vinding-Diers

Known for: Pinot Noir

Signature wine: Treinta y Dos

Other labels: Barda, Mainqué

Visiting: Open to the public by appointment only; tasting room

Chacra—owned by Incisa della Rocchetta, whose family owns Sassicaia, and two silent partners—is the alter ego of Patagonia's renowned Bodega Noemía. The two estates are neighbors, sharing the 1955 and 1932 vineyards, which are planted with old field blends of Malbec and Pinot Noir, and their winemaker, Hans Vinding-Diers. From these older vineyards, Chacra uses only the Pinot Noir, and the wines are named for the dates of the vineyards: "Cincuenta y Cinco" (from a vineyard planted in 1955), "Sesenta y Siete" (from 1967), and "Treinta y Dos" (from all the way back to 1932). The wines are 100 percent biodynamically produced and grown and are frequently considered the grands crus of Argentina's Pinot Noir.

Chakana

Founded: 2002

Address: Ruta Provincial 15, km 34, Luján de Cuyo, Mendoza, www.chakanawines .com.ar

Owners: Pelizzatti family, Fenix Partners Fund

Winemaker: Gabriel Bloise

Consulting winemaker: Alberto Antonini

Viticulturists: Alan York, Facundo Bonamaizon

Known for: Malbec

Signature wine: Ayni

Other labels: Maipe, Cueva de las Manos, Nuna

Visiting: Open to the public by appointment only; tasting room

Chakana, taking its name from the local Indian word for the Southern Cross, is located approximately twenty miles south of Mendoza City. In addition to working with Alberto Antonini, Chakana initially consulted with Chile's Pedro Parra about the optimal match of soils, rootstocks, and varieties. The wines are made from grapes grown on 371 acres in Agrelo, 52 more in Mayor Drummond, and 105 in Altamira.

Chandon Argentina

Founded: 1959

Address: Ruta Provincial 15, km 29, Agrelo, Luján de Cuyo, Mendoza, www.chandon.com.ar

Owner: LVMH

Winemakers: Onofre Arcos, Hector Agostini

Viticulturist: Martín Reboredo

Known for: Chardonnay, Pinot Noir

Signature wine: Chandon Signature Brut

Other labels: Baron B., Latitud 33, Valmot, Mercier, O2, Castel

Visiting: Open to the public by appointment only; tasting room; restaurant by appointment only

In the late 1950s, Renaud Poirier, a well-known French enologist and Claude Moët's technical adviser, conducted several enological trials and was convinced that Agrelo was the ideal place to locate Moët & Chandon's first winery outside France. Uco Valley is the source of its high-end wines. The winery makes spar-kling wines using both the *méthode traditionnelle* and the Charmat method. The winery restaurant, L'Orangerie, does a great sparkling wine–paired lunch by appointment.

Cheval des Andes

Founded: 1999

Address: Las Compuertas 5549, Luján de Cuyo, Mendoza, www.chevaldesandes .com

Owners: LVMH (50%), Château Cheval Blanc (50%)

Winemakers: Nicolas Audebert, Pierre Lurton, Pierre Oliver Clouet

Consulting winemaker: Gilles Pauquet

Viticulturist: Gustavo Ursomarso

Known for: Malbec–Cabernet Sauvignon blends, Petit Verdot, Merlot, Cabernet Franc

Signature wine: Cheval des Andes

Visiting: Open to the public by appointment only; tasting room

Cheval des Andes is a high-end wine collaboration between LVMH's Terrazas de los Andes and Pierre Lurton of Cheval Blanc. Pierre proposed the joint venture in 1999 when he realized the potential of the great old-vine sites in Mendoza. Cheval comes from around eighty-nine acres of vineyards, two-thirds located in Vistalba and the balance in Uco Valley's La Consulta. Nicolas Audebert has an impressive CV, including spells at Veuve Clicquot, Krug, and, of course, Château Cheval Blanc. He also is a big polo fan and built a regulation-size field at the winery.

Clos de los Siete

Founded: 1998

Address: Clodomiro Silva sin número, Vista Flores, Tunuyán, Mendoza, www.clos7.com.ar

Owners: Monteviejo, Cuvelier Los Andes, Bodega Rolland, DiamAndes

Winemaker: Michel Rolland

Viticulturist: Carlos Tizio

Known for: Malbec, Cabernet Sauvignon, Merlot, Syrah, Petit Verdot

Signature wine: Clos de los Siete

Visiting: Open to the public by appointment only

The visionary collaborative enterprise of Clos de los Siete is the dream of Michel Rolland, who, starting in 1998, brought together seven adjacent winery projects in Tunuyán (his own and six others), planted on 2,125 acres of desert in the Uco Valley. It encompasses nine vineyard parcels, including the Altamira property at 3,124 feet (owned by Bernard Cuvelier), Lindaflor and Monteviejo (the late Catherine Péré-Vergé), Altaflor (a jointly owned venture), Viña los Dassos (Laurent Dassault's Flechas de los Andes), Herbaflor (Rothschild), Primaflor and Rocaflor, the basis of the Alta Vista wines (the d'Aulan family), and Mariflor, at 4,000 feet (Michel Rolland). Altavista decided in the end to go it alone. Of the other six wineries planned, three have been constructed, and each is an exercise in architectural expression. Starting with a blank slate (and, no doubt, blank checks from the owners), the winemaking facilities of Clos de los Siete are among the world's most sophisticated. Clos de los Siete is a flagship blend of red

grapes from all the estates on the property, made by Michel Rolland, and makes up approximately 80 percent of the total output of all the wineries. There is great excitement surrounding the release of a complementary white wine in 2014.

Colomé

Founded: 1831

Address: Ruta Provincial 53, km 20, Molinos, Salta, www.bodegacolome.com

Owner: Hess family estate

Winemaker: Thibaut Delmotte

Consulting winemaker: Randle Johnson (Hess family winemaker)

Viticulturist: Javier Grané

Known for: Malbec, Torrontés

Signature wine: Colomé Malbec Reserva

Visiting: Open to the public; tasting room; restaurant and inn (Estancia Colomé)

Colomé is one of the oldest official wineries in Argentina, established in 1831, and also one of the world's highest vineyards, at close to ten thousand feet. The estate encompasses 96,000 acres, including 32 acres of vineyards that are over 150 years old. Hess owns two other vineyards in Salta: one in Payogasta, focused on reds, and one in El Arenal–Río Blanco, which is even higher than Colomé. Since 2001, Colomé has been biodynamically farmed. It has an amazing art museum (designed by the artist James Turrell), a visitor center and restaurant, and a small hotel in addition to the beautiful Hess winery, which was completed in 2006.

Cruzat

Founded: 2004

Address: Costa Flores sin número, Perdriel, Mendoza, www.bodegacruzat.com

Owners: Carlos Barros, Gastón Cruzat, Hernán Boher, Pedro Rosell

Winemaker: Pedro Rosell

Viticulturist: Santiago Mayorga

Known for: Pinot Noir, Chardonnay

Signature wine: Cruzat Cuvée Réserve

Visiting: Open to the public; tasting room

This hidden jewel of a winery eluded me on several previous visits to Argentina. It makes nothing but sparkling wine, and it does it very well. Pedro Rosell is an enthusiastic, energetic almost-octogenarian and one of the true maestros of sparkling wine in South America, in the same echelon as Brazil's Mario Geisse. The partners are committed to achieving greatness in effervescence and vinifying *méthode traditionnelle* wines exclusively from Pinot Noir and Chardonnay.

Cuvelier Los Andes

Founded: 2003

Address: Clodomiro Silva sin número, Vista Flores, Tunuyán, Mendoza, www.cuvelierlosandes.com

Owner: Cuvelier family

Winemaker: Andrian Manchon

Consulting winemaker: Michel Rolland

Viticulturist: Carlos Tizio

Known for: Malbec

Signature wine: Grand Vin

Other labels: CLA Colección

Visiting: Open to the public; tasting room

The Cuvelier winery is one of the participants in the Clos de los Siete project and also owns Château Léoville-Poyferré, a classified second growth in Saint-Julien, Bordeaux. With 170 acres of vines, it produces three wines: a Grand Vin, a Grand Malbec, and a tasty second label, Colección. Unlike most of the other Clos de los Siete wineries, Cuvelier is open to visitors, offering tastings of the different grape varieties, horseback riding on the 850-acre farm, and bicycle riding for the more courageous.

Decero

Founded: 2000

Address: Bajo las Cumbres 9003, Agrelo, Luján de Cuyo, Mendoza, www.decero.com

Owner: Thomas Schmidheiny

Winemaker: Marcos Fernández

Consulting winemaker: Paul Hobbs

Viticulturist: Matías Cano

Known for: Malbec, Cabernet Sauvignon, Petit Verdot

Signature wines: Mini Ediciones Petit Verdot, Decero Malbec, Amano

Visiting: Open to the public by appointment only; tasting room; restaurant (Finca Decero)

Appropriately named Decero, meaning "from scratch," this winery was born of a bare patch of land and a family's love affair with wine. The Swiss Schmidheiny family has worldwide connections to wine. Just as a visit to Napa Valley inspired Thomas's mother, Adda, in the 1970s, Agrelo captured his heart and mind when he traveled to Argentina twenty years later. All wines come from the 272-acre Remolinos vineyard in Alto Agrelo, named for the tiny whirlwinds that thread their way along the vines, keeping the grapes dry and in perfect condition. The "Mini Ediciones" Petit Verdot is Decero's most distinctive wine and one of the best pure examples of the grape I have enjoyed.

Desierto

Founded: 2003

Address: Ruta Provincial 34, km 2, 25 de Mayo, La Pampa, www.bodegadel desierto.com.ar

Owners: Armando and María Loson

Winemaker: Sebastián Cavagnaro

Consulting winemaker: Paul Hobbs

Viticulturist: Enzo Mugnani

Known for: Cabernet Franc, Malbec, Merlot, Cabernet Sauvignon, Syrah, Sauvignon Blanc, Chardonnay, Pinot Noir

Signature wines: Desierto 25/5, Pampa

Visiting: Open to the public by appointment only

Bodega del Desierto is the first winery in La Pampa. Located in the province's appellation of Alto Valle Rio Colorado near the town of 25 de Mayo, the winery, as the name suggests, is set in the middle of nowhere in the dry cactus desert, a good two-hour drive from Neuquén. It encompasses 445 acres of vineyard, all under irrigation. The owners claim that they established the winery here because it was as good a spot as any (and, I suspect, because land is comparatively cheap). The wines suggest that they may be onto something.

DiamAndes

Founded: 2005

Address: Clodomiro Silva sin número, Vista Flores, Tunuyán, Mendoza, www.diamandes.com

Owner: Bonnie family

Winemaker: Silvio Alberto

Consulting winemaker: Michel Rolland

Known for: Malbec

Signature wine: DiamAndes de Uco Gran Reserva

Other labels: Perlita, L'Argentin de Malartic

Visiting: Open to the public by appointment only; tasting room; restaurant

Another member of the Clos de los Siete group, this winery is owned by the doyens of Château Malartic-Lagravière and Château Gazin Rocquencourt in the Graves region of Bordeaux. The name is a pun on the words *diamante* (Spanish for "diamond") and *Andes* that the owners came up with one day while enjoying an *asado* at Laguna del Diamante in Mendoza and admiring the reflection of the Maipo volcano on the lake. The stunning architecture of the winery is the work of the renowned Mendocino architects Mario Yanzón and Eliana Bórmida. Completed in 2009, it is geared up for visitors, complete with a seated tasting room and a wine bar.

Domingo Molina (Domingo Hermanos)

Founded: 1999

Address: Yacochuya Norte sin número,

Cafayate, Salta, www.domingomolina
.com.ar

Owner: Domingo family

Winemaker: Rafael Domingo

Viticulturist: Osvaldo Domingo

Known for: Malbec, Torrontés, Tannat

Signature wine: Rupestre

Visiting: Open to the public by appointment only; tasting room

As with many Argentinean wine companies, there's more to Domingo Molina than meets the eye. It produces two separate and distinctive brands, Hermanos and Rupestre. It is a small, new winery with a panoramic view over a hillside covered in cacti, dry rock and sand, and vines. Domingo Molina is a family winery managed by three brothers (Osvaldo, Gabriel, and Rafael), representing the second generation of Domingos making wine together (hence the brand name Hermanos, Spanish for "brothers"). *Rupestre* means "cave drawings," and symbols on the label are derived from the thousand-year-old drawings in caves located on the property. The Rupestre vineyard was one of the first to be planted in the region, at 7,380 feet above sea level.

Dominio del Plata Winery

Founded: 1999

Address: Cochabamba 7801,
    Agrelo, Luján de Cuyo, Mendoza,
    www.dominiodelplata.com.ar

Owner: Susana Balbo

Winemakers: Susana Balbo, José Lovaglio

Viticulturist: Alejandro Eaton

Known for: Malbec, Cabernet Sauvignon,
    Cabernet Franc, Petit Verdot, Merlot,
    Tannat, Syrah, Bonarda, Torrontés

Signature wines: Susana Balbo Malbec,
    Crios Torrontés

Other labels: Crios, Benmarco, Susana
    Balbo, Nosotros, Bodini, Zohar, Anubis

Visiting: Open to the public by appointment only; tasting room; restaurant
    (Osadía de Crear)

The wine world is fortunate that Susana Balbo's family refused to let her study nuclear physics, her original calling. Opting for wine instead, she became the first Argentinean woman to qualify as an enologist. Since then she has earned a reputation as one of the finest winemakers in the world. She started at Michel Torino and moved to Catena before establishing her own Dominio del Plata. She is a tireless perfectionist who won't take no for answer. When she couldn't get the equipment she needed in the early 1990s, she simply went to California and packed it in her luggage—everything from a refractometer to a fax machine! Her purity of style is reflected in site-specific and varietally spot-on wines. The Benmarco line comprises her premium offering, while the Crios wines are proverbial great values. In May 2013, Edgardo del Pópolo, formerly of Doña Paula, joined the winery as chief operating officer, a move that will surely pay dividends. Also in 2013, the Osadía de Crear restaurant opened at the winery, a project of chef José Cacciavillani and his wife, Ana.

Doña Paula

Founded: 1997

Address: Poliducto YPF sin número,

Ugarteche, Luján de Cuyo, Mendoza, www.donapaula.com

Owner: Santa Rita Group

Winemaker: David Bonomi

Consulting viticulturist: Edgardo del Pópolo

Known for: Malbec, Cabernet Sauvignon, Chardonnay, Sauvignon Blanc, Torrontés

Signature wine: Doña Paula Selección de la Bodega Malbec

Other labels: Paula, Los Cardos

Visiting: Open to the public by appointment only

Doña Paula is the foundation of the Chilean Claro Group's efforts in Argentina. Claro (also known as the Santa Rita Group) already owned the Chilean producers Santa Rita and Carmen and had a partnership with Domaines Barons de Rothschild (Lafite) in Los Vascos. When they wanted in on Argentina, they were fortunate to connect with the talented Edgardo del Pópolo (Edy). Edy now serves in a purely consultative role. The facility is situated in an old vermouth plant converted to fine-wine production. The operation has several vineyards, with the primary, 1,060-acre estate located in Ugarteche. Despite its size, it is committed to using estate-grown fruit; 100 percent of the wines are made from estate grapes (except for their Torrontés, from Cafayate). The Los Cardos wines are readily available and among the world's great values.

Durigutti

Founded: 2003

Address: Roque Sáenz Peña 8450, Las

Compuertas, Luján de Cuyo, Mendoza, www.durigutti.com

Owners: Héctor Durigutti, Pablo Durigutti

Winemakers: Héctor Durigutti, Pablo Durigutti

Viticulturist: Héctor Durigutti

Known for: Malbec

Signature wine: Malbec Clásico

Other labels: Lamadrid

Visiting: Open to the public by appointment only

The Durigutti brothers are one of the continent's more formidable sibling teams. They craft three small-production Malbecs and two of the country's best examples of Bonarda. Héctor, considered one of Argentina's top winemakers, first became known for his work alongside Alberto Antonini in founding Altos Las Hormigas. With extensive experience in Europe (including a stint at Antinori), Héctor has developed a unique winemaking style that blends tradition with innovation. Héctor's younger brother, Pablo, has also built a star-studded résumé including experience with top names such as Rutini, Furlotti, Catena, and Cinco Tierras. Today both brothers are among the country's most sought-after wine consultants. Héctor's style is considered more European and Pablo's more American; the combination makes for a great synergy.

Furlotti: Pablo also makes wine for Bodega Furlotti, a winery whose history dates back to 1889, when Angel Furlotti arrived in Argentina from Parma. In 1917, his white and red wines won awards in Milan. By 1921, they were being sold in Buenos Aires and other provinces of Argentina.

The bodega is a project of Angel Furlotti's great-granddaughter Gabriela Furlotti.

Lamadrid: Héctor is winemaker for Lamadrid (www.lamadridwines.com), owned by Guillermo García Lamadrid, with three vineyards in Agrelo: Finca La Matilde (planted in 1929), Finca Lamadrid (1973), and Finca La Suiza (2004). This winery makes single-estate wines with only indigenous yeasts and without fining or filtration.

Enrique Foster

Founded: 2002

Address: San Martín 5039, Carrodilla, Luján de Cuyo, Mendoza, www.bodega foster.com

Owner: Enrique Foster Gittes

Winemaker: Mauricio Lorca

Viticulturist: Enzo Mugnani

Known for: Exclusively Malbec

Signature wine: Enrique Foster Limited Edition

Other labels: Firmado, Terruño, Ique, Foster Pink

Visiting: Open to the public by appointment only

The Enrique Foster winery is the first purpose-built winery in the world to produce only Malbec varietal wines by gravity flow. It has two vineyards in Luján de Cuyo: thirty acres planted in 1919 in Carrodilla, where the winery is located, and another twenty in Las Compuertas, planted in 1966. Mauricio Lorca has been with Enrique Foster for a number of years. Before that he perfected his craft at Catena (Bodegas Esmeralda), Michel Torino, Finca La Celia, and Luigi Bosca. He still advises close to twenty wineries and also makes wine under his own label (www.mauricio lorca.com), sourcing fruit from Vistaflores.

Ernesto Catena Vineyards

Founded: 1998

Address: Ruta Provincial 92 sin número, Vista Flores, Tunuyán, Mendoza, www.ernestocatenavineyards.com, www.almanegrawines.com.ar, www.siestawines.com.ar

Owner: Ernesto Catena

Winemaker: Alejandro Kuschnaroff

Viticulturist: Alejandro Kuschnaroff

Known for: Malbec, Cabernet Sauvignon, Syrah, Pinot Noir, Cabernet Franc, Petit Verdot, Viognier

Signature wines: Siesta, Alma Negra

Visiting: Open to the public by appointment only; tasting room

Ernesto Catena, son of Nicolás and brother of Laura, must be one of the hardest-working winemakers in Argentina. Ernesto has degrees in computer science, economics, design, and history and owns Argentina's leading modern photography gallery, in the Palermo neighborhood of Buenos Aires. He sources different varieties of grapes from diverse vineyards, working with small growers to obtain the best fruit. This means that all of his wines have distinctive identities, including the densely concentrated Siesta wines and the sumptuous Alma Negra *cortes.*

Fabre Montmayou

Founded: 1992

Address: Roque Saenz Peña sin número,

Vistalba, Luján de Cuyo, Mendoza,
www.domainevistalba.com

Owner: Hervé Joyaux Fabre

Winemaker: Matías Riccitelli

Consulting winemaker: Paul Hobbs

Known for: Malbec

Signature wine: Grand Vin

Other labels: Phebus

Visiting: Open to the public by appointment only; tasting room; restaurant

Don't let the website address confuse you.
The independent winery is the property of
the Frenchman Joyaux Fabre, the son of
Bordelaise wine *négociants,* who came to
Argentina in 1990 and, after some search-
ing, acquired a château-style property surrounded by vineyards, including a parcel
of old head-trained Malbec vines dating
back to 1908. Today he owns close to 220
acres of old vines centered in Vistalba and
another 120 in the Upper Río Negro Valley,
Patagonia, where he has a second winery.

Familia Arizu

See Luigi Bosca, Viña Alicia

Familia Cecchin

Founded: 1959

Address: Manuel A. Sáenz 626, Russell,
Maipú, Mendoza, www.bodegacecchin
.com.ar

Owner: Oscar Alberto Cecchin

Winemaker: Carlos Fernández

Consulting winemakers: Emile Heredia,
Vincent Wallard

Viticulturist: Adriana Carrion

Known for: Malbec, Carignan

Signature wine: Blend de Familia

Visiting: Open to the public; tasting room;
restaurant (A la Sombra)

The Cecchin family pride themselves on
the fact that all their wines are produced
from certified organic grapes grown on
their 27-acre estate. A third-generation
wine family, the Cecchins work in a very
traditional manner: they use horse-drawn
plows, minimize the use of agricultural
chemicals, and use only native yeasts.
The vineyards are bordered with aromatic
plants like *te de burro* and *amaranto* to attract animal pests and keep them away
from the vines. Their strong focus on
Carignan is unique. The Cecchins also
make their own organic olive oil and cultivate fruit and walnuts that are used in
their restaurant.

Familia Schroeder

Founded: 2001

Address: Calle 7 Norte, San
Patricio del Chañar, Neuquén,
www.familiaschroeder.com

Owner: Roberto Schroeder

Winemaker: Leonardo Puppato

Consulting winemaker: Paul Hobbs

Viticulturist: Marcelo García

Known for: Pinot Noir, Malbec, Cabernet
Sauvignon

Signature wine: Familia Schroeder Pinot
Noir–Malbec

Other labels: Alpataco, Saurus

Annual production 1.2 million bottles,
100% estate grown

Visiting: Open to the public; tasting room; restaurant (Saurus)

One of the first things you notice at the Familia Schroeder winery is a museum housing the fossil of a large dinosaur. When excavating the land for the winery, the builder unearthed the petrified bones of a forty-foot-long dinosaur that proved to be about seventy-five million years old. Its discovery was the inspiration for the Saurus line of wines. Together the wines, the dinosaur, and the excellent restaurant attract over twenty thousand tourists each year. The winery is built on a slope with a 72-foot differential over five levels, a key feature that makes it one of the few true gravity-flow wineries in Argentina. Familia Schroeder was the first winery in Argentina to achieve ISO Certification 22.000 (for food safety and traceability) throughout the entire chain of production. Its wines are very classic of the Neuquén region, and as such more about elegance than power.

Familia Zuccardi

Founded: 1963
Address: Ruta Provincial 33, km 7.5, Maipú, Mendoza, www.santajulia.com.ar, www.familiazuccardi.com
Owner: José Alberto Zuccardi
Winemaker: Rodolfo Montenegro
Viticulturists: Edgardo Consoli, Sebastián Zuccardi
Known for: Malbec, Cabernet Sauvignon, Tempranillo, Bonarda, Torrontés
Signature wines: Santa Julia Reserva Malbec, Zuccardi Zeta

Other labels: Zuccardi, Santa Julia, Chimango (organic line)
Visiting: Open to the public; tasting room; restaurants (Casa del Visitante, Pan & Oliva)

If you can visit only a couple of wineries in Mendoza, I would recommend that Zuccardi be one of them. Aside from having gold-standard hospitality and two superb restaurants, the winery makes a dizzying array of wines in multiple varieties, prices, and styles. There's something for everyone—Caladoc to Cabernet, Marselan to Malbec—made under the supervision of the passionate and ever-experimenting patriarch, José Alberto. The estate has a small experimental winery with twenty stainless-steel tanks of varying sizes. The objective is to experiment with up to twenty-five new varieties over five years, along with trying out new vinification techniques to determine the various grapes' suitability for commercialization and the best vinification techniques for each variety. The winery tours, led by well-trained staff, are excellent and thorough.

Finca Dinamia

Founded: 2002
Address: 1545 Hipólito Yrigoyen Ave., San Rafael, Mendoza, fincadinamia.com
Owner: Alejandro Bianchi
Winemaker: Fabricio Orlando
Consulting winemaker: Mauro Nosenzo
Viticulturist: Pablo Miguelo
Known for: First Malbec in Argentina to receive Demeter organic certification
Signature wine: Buenalma

Visiting: Open to the public; two-bedroom guesthouse available by reservation

This recent arrival is the dream of a cousin of the Bianchi family, who operate the eponymous winery, and its former chief marketing officer. Now under construction, the winery aims to establish the first entirely organic and biodynamic vineyard in Argentina. The vineyards on the 62-acre property are planted wholly to Malbec, with the rest of the land dedicated to a fruit orchard and small animal farm. The winery has an interesting vine-adoption program, offering personalized wines from the harvest (see www.mivin-iabio.com).

Finca Flichman

Founded: 1910

Address: Munives 800, Barrancas, Maipú, Mendoza, www.flichman.com.ar

Owner: Sogrape Vinhos SA

Winemaker: German Berra

Viticulturist: Cecilia Acosta

Known for: Cabernet Sauvignon, Malbec, Shiraz, Chardonnay, Merlot, Sauvignon Blanc

Signature wines: Dedicado, Caballero de la Cepa

Other labels: Paisaje, Expresiones, Gestos, Misterio, Roble

Visiting: Open to the public; tasting room

Finca Flichman was established in 1873 by the Polish Jewish textile merchant Sami Flichman. He and his family grew and sold grapes before starting their own winery over thirty years later. The prominent Guedes family, owners of Portugal's Sogrape, purchased the winery in 1998 and have invested in upgrading both the vineyards and the winery while expanding production and exports. With some 2,350 acres in Maipú and Tupungato, Flichman grows 890 acres of *vinifera*, with the balance planted to hybrid and *labrusca* grapes. The winery created one of the earliest super-premium wines, "Caballero de la Cepa," an iconic bottling that remains popular today.

Finca La Celia

Founded: 2000

Address: Circunvalación Celia Bustos de Quiroga 374, Eugenio Bustos, San Carlos, Mendoza, www.fincalacelia .com.ar

Owner: VSPT Wine Group Argentina

Winemaker: Cristian García

Consulting winemaker: Paul Hobbs

Viticulturist: Manuel Bianchi

Known for: Malbec, Cabernet Franc, Tannat, Petit Verdot, Chardonnay

Signature wine: La Celia Supremo

Other labels: La Consulta, Furia, La Finca, Angaro, Eagle's Rock

Visiting: Open to the public by appointment only; tasting room; restaurant (Posada La Celia); inn

In 1882, Eugenio Bustos decided to move south of the river Tunuyán in order to develop a Malbec-focused estate. The subsequent winery was named La Celia in honor of his daughter, Celia. When she inherited the property, she expanded the operation, which was ultimately acquired by the VSPT Wine Group. A well-appointed five-room inn sits on the property, with a lovely restaurant. The winery offers tourist packages that let you blend your own wine, participate in the harvest, work on off-season pruning, bike around the property, and take cooking classes. It also offers a venue for meetings.

Finca Las Nubes

Founded: 1998
Address: El Divisadero, km 4, Cafayate, Salta, www.bodegamounier.com
Owner: José Luis Mounier
Winemaker: José Luis Mounier
Known for: Torrontés, Malbec, Tannat
Signature wine: Finca Las Nubes Torrontés
Visiting: Open to the public; tasting room; restaurant

José Luis Mounier, the founder of Finca Las Nubes, is universally considered the maestro of Torrontés. After working with the Etchart family, where he was winemaker for two decades, he and his wife decided to buy their own property and build a winery to pass on to their children. They identified a property at 4,400 feet and named it Finca Las Nubes, "farm of the clouds." The certified organic winery sells over half of its wines direct and, if you book ahead, can offer lunch.

Finca Sophenia

Founded: 1997
Address: Ruta Provincial 89, km 12.5, Gualtallary, Tupungato, Mendoza, www.sophenia.com.ar
Owners: Roberto Luka, Gustavo Benvenuto
Winemaker: Rogelio Rabino

Consulting winemakers: Matías Michelini, Michel Rolland
Viticulturist: Rogelio Rabino
Known for: Malbec, Viognier, Cabernet Sauvignon, Merlot, Chardonnay, Sauvignon Blanc
Signature wine: Sophenia Synthesis the Blend
Other labels: Altosur, E.S. Vino
Visiting: Open to the public by appointment only; tasting room

Roberto Luka, principal owner of the winery, is a former president of Wines of Argentina. Named for the owners' two daughters, Sophia and Eugenia, the property has a highly capable production team, led by Rogelio Rabino and assisted by the rock-star consultants Matías Michelini and Michel Rolland. All of their wines are well made and noteworthy, but they make one of the most striking Sauvignon Blancs in all of Argentina.

Flechas de los Andes

Founded: 2003
Address: Clodomiro Silva sin número, Vista Flores, Tunuyán, Mendoza, www.flechasdelosandes.com.ar
Owners: Compagnie Vinicole Edmond de Rothschild, Laurent Dassault
Winemaker: Pablo Richardi
Consultant: Michel Roland
Viticulturist: Marcelo Canatella
Known for: Malbec
Signature wine: Flechas de los Andes Gran Corte
Visiting: Tasting room

Another participant in the Clos de los Siete project, Flechas de los Andes, meaning "arrows of the Andes," is a partnership between Laurent Dassault, owner of Château Dassault in Saint-Émilion, and Baron Benjamin de Rothschild, a Swiss banker and the chairman of the Edmond de Rothschild Group. Perhaps tellingly, Dassault's grandfather founded the legendary Dassault aircraft factory, some of whose aircraft and missiles aided the Argentinean Air Force during the Falklands War. The winery, designed by Phillip Duillet (art director for the Star Wars films), makes use of myriad symbols of arrows throughout the gravity-flow winery.

Goulart

Founded: 1997

Address: Ruta 50 Vieja, Rodeo del Medio, Mendoza, www.fincalugildegoulart .com.ar

Owner: Erika Goulart

Winemaker: Erika Goulart

Consulting winemaker: Alejandro Canova

Viticulturist: Marcelo Canatela

Known for: Malbec

Signature wine: Goulart Grand Vin Malbec

Visiting: Open to the public by appointment only

The vivacious Brazilian Erika Goulart came to the wine business by accident. Married to an Argentinean, she is the granddaughter of Marshal Gastão Goulart, the leader of Brazil's constitutional revolution in 1932. He purchased the land in Argentina (seventy acres in Lun-

lunta) in 1915. In 1988, Erika discovered the deed to the properties in Mendoza, moved her family to Argentina, reclaimed the vineyards, and started the winery operation, doing major replanting and vineyard work over six years. Among the winery's holdings are the Don Pedro vineyard (named for Erika's eldest son), the Don Lucas vineyard (named for her youngest son), and the Don José vineyard (named for her husband).

Graffigna

Founded: 1870

Address: Colón 1342 Norte, Desamparados, San Juan, www.graffignawines.com

Owner: Pernod Ricard Argentina

Winemaker: Gerardo Dantiz

Consultant: Phil Laffer

Viticulturist: Mariano Bustos

Known for: Malbec, Cabernet Sauvignon, Pinot Grigio, Syrah, Chardonnay

Signature wine: Centenario Malbec

Visiting: Open to the public by appointment only

The Graffigna winery is named for one of Argentina's pioneering enologists. In 1923, Santiago Graffigna established the country's first gravity-flow winemaking operation. By 1925, the winery had imported more than eight hundred varieties of grapes to study, more than any other winery in the country. Graffigna has been a leader in both San Juan and the Pedernal Valley. Purchased by Allied Domecq in 1980, the winery was acquired by Pernod Ricard after Allied's dissolution in 2005.

Grupo Peñaflor

See Las Moras, Michel Torino, Trapiche

Owned by the Bemberg family since 2010, Grupo Peñaflor is one of the world's largest wine companies, with wineries all across the wine regions of Argentina, each operating independently and successfully. While Las Moras, Michel Torino, and Trapiche are the best known, the company also owns the following notable wineries:

Andean Viñas: An export-focused effort based in Luján de Cuyo, this brand produces a wide range of wines with an excellent price-to-value ratio.

Santa Ana: This winery has been successful since the 1970s. Its "Unanime" was named the best Argentinean wine in *Decanter*'s 2013 feature "58 Great Value New World Reds."

Humberto Canale

Founded: 1909
Address: Chacra 186, General Roca, Río Negro, www.bodegahcanale.com
Owner: Guillermo A. Barzi Canale
Winemaker: Horacio B. Bibiloni
Consulting winemakers: Susana Balbo, Pedro Marchevsky
Viticulturist: Juan Martín Vidiri
Known for: Pinot Noir, Merlot, Malbec
Signature wine: H Canale Centenium
Other labels: Intimo, Canale Black River, Diego Murillo
Visiting: Open to the public; tasting room

The Canales can be called the first family of Patagonia. Over one hundred years old, the winery is closely connected to Argentina's development of Patagonia, a project realized with the construction of the railroad. The founding Canale, an engineer on the project, was convinced that Patagonia was a great wine region. After traveling to France, he planted imported Bordelaise grape varieties. The winery is still run by the Canale family: Barzi is the founder's great-nephew. Many of the newer wineries in the region were inspired by the longevity and success of the Canales.

Kaiken

Founded: 2001
Address: Roque Saenz Peña 5516, Vistalba, Mendoza, www.kaikenwines.com
Owner: Aurelio Montes
Winemaker: Aurelio Montes Jr.
Viticulturist: Aurelio Montes Jr.
Known for: Malbec
Signature wine: Mai
Other labels: Terroir Series, Ultra
Visiting: Open to the public; tasting room; restaurant

Kaiken is the brainchild of Chile's Aurelio Montes, who went to Argentina because of its grapes, in partnership with the late Douglas Murray. Kaiken was founded on a quest to find the best grapes in Mendoza, rather than to get involved with big investments in land and a winery. This approach has enabled Kaiken to tap into all the quality-producing areas in Mendoza by entering into contracts with top growers, closely supervising the viticulture, and insisting on low yields that are also closely controlled. Like some California wineries, Kaiken pays growers considerably above market price so that it can take its pick of grapes. A *kaiken* is a wild goose that flies over greater Patagonia, traveling between southern Argentina and Chile, much like Montes.

Krontiras

Founded: 2004
Address: Olavarría 4900, Perdriel, Luján de Cuyo, Mendoza, www.bodegaskrontiras.com
Owner: Constantino Krontiras
Winemaker: Leonardo Sesto
Consulting winemaker: Panos Zoumboulis
Viticulturist: Maricruz Antolín
Known for: Malbec
Signature wine: Doña Silvina Reserve Malbec
Visiting: Open to the public by appointment only

It may seem odd to have a Greek-named winery in South America, but its owner, Constantino Krontiras, was born in Greece. He is married to Silvina Macipe, a Mendoza native. The Doña Silvina Reserva Malbec, coming from a ninety-year-old vineyard, is named for her. The winemaker, too, is Greek: Panos Zoumboulis brings more than twenty-five years' experience of Old World wine production to this project in Argentina. The winery is certified organic and will soon be a Demeter-certified biodynamic property that embraces viticulture as a living system. Krontiros makes olive oil and grappa in addition to four wines.

Lagarde

Founded: 1897
Address: San Martín 1745, Luján de Cuyo, Mendoza, www.lagarde.com.ar
Owner: Sofia Pescarmona
Winemaker: Juan Roby Stordeur
Viticulturist: Juan Roby Stordeur
Known for: Sauvignon Blanc, Viogner, Chardonnay, Muscat, Malbec, Merlot, Syrah, Cabernet Sauvignon, Cabernet Franc
Signature wine: Henry Gran Guarda No. 1
Other labels: Sémillon 1942, Henry, Primeras Viñas, Guarda, Altas Cumbres
Visiting: Open to the public; tasting room; restaurant (Lagarde)

Lagarde is one of the oldest and most traditional wineries in Mendoza. The third generation of the Pescarmona family is in charge; the family purchased the winery in 1977 from the descendants of the Pereira family, who founded the bodega. Lagarde is a member and founder of the committee that established the Luján de Cuyo controlled appellation of origin. They work from six vineyards, four in Luján and two in the Uco Valley. Their Gran Guarda, a red *corte,* is among Argentina's most consistent super-premium red wines. A delightful five-course tasting lunch is served in the nineteenth-century farmhouse.

Lamadrid. *See* Durigutti

La Posta

Founded: 2002
Address: Ruta 92 sin número,

Vista Flores, Tunuyán, Mendoza,
www.lapostavineyards.com

Owner: Laura Catena

Winemaker: Luis Reginato

Known for: Malbec

Signature wine: Angel Paulucci Vineyard
Malbec

Visiting: Open to the public by appointment only

The La Posta wines are a joint venture
between Laura Catena and her U.S. importer, Vine Connections. La Posta makes
single-vineyard wines from selected old-
vine plantings throughout the Uco Valley, each of which have been identified
for their depth of character, the history of
their land, and the personality of the families who own them.

La Riojana

Founded: 1940

Address: La Plata 646, Chilecito, La Rioja,
www.lariojanawines.com

Owner: Board of members

Winemaker: Rodolfo Griguol

Viticulturist: Viviana Michel

Known for: Malbec, Torrontés

Signature wine: RAZA Malbec Reserva

Other labels: Ecologica, Solombra, Viñedos
de la Posada, Fair & Square, Tilimuqui,
Santa Florentina

Visiting: Open to the public; tasting room

The huge La Riojana co-op is renowned
in Argentina not only for being the largest producer in La Rioja but also for its
ground-breaking clonal work and yeast
trials on Torrontés, which has benefited

the entire Argentinean wine industry.
Originally an agricultural cooperative,
the wine co-op split from the umbrella
operation in 1989. It works with close to
five hundred growers, representing about
80 percent of the region and nine thousand acres, mostly in the Famatina Valley. Certified organic in 2000, La Riojana
has become one of world's largest producers of organic wine: seven hundred acres
are now certified organic, and that area
is growing.

Las Moras

Founded: 2001

Address: Rawson sin número, San Martín,
San Juan, www.fincalasmoras.com

Owner: Grupo Peñaflor

Winemaker: Eduardo Casademont

Viticulturist: Claudio Rodriguez

Consulting winemakers: Michel Rolland,
Richard Smart, Daniel Pi

Known for: Malbec, Shiraz, Viognier,
Sauvignon Blanc

Signature wine: Mora Negra

Visiting: Open to the public; tasting room

A leader in the Pedernal Valley, Las Moras
is named for the fruit of the *moreras* (mulberry trees) that surround the property. It
is said that if the mulberries are good in
a particular year, the grapes will be too.
Though the winery itself is located in
Tulum, its leading vineyard estate is located in the newer, high-altitude region.
And while it makes a range of wines, it is
acknowledged to be one of the leading producers of Syrah in Argentina.

Luca Winery

Founded: 1999

Address: Calle Cobos sin número, Agrelo, Luján de Cuyo, Mendoza, www.lucawines.com

Owner: Laura Catena

Winemaker: Luis Reginato

Known for: Malbec, Syrah, Pinot Noir, Chardonnay

Signature wine: Nico

Visiting: Open to the public by appointment only

Laura Catena is a San Francisco–based emergency room surgeon, educated at Harvard and Stanford; a fourth-generation vintner; an amazing tango dancer; and the author of a very good book on Argentinean wine and culture, *Vino Argentino*. It is a must read if you plan to travel there. Wines at Luca, named for her eldest son, are made from older Uco Valley vineyard sites; she owns no vineyards. The top wine, Nico by Luca, is an impressive Malbec blend of two historic old-vine vineyards, Rosas and Paganotto. Luis Reginato, the winemaker, also oversees the vineyards for the Catena Zapata wineries.

Luigi Bosca—Familia Arizu

Founded: 1901

Address: San Martín 2044, Mayor Drummond, Luján de Cuyo, Mendoza, www.luigibosca.com.ar

Owner: Arizu family

Winemaker: José Irrera, Vicente Garzia

Viticulturist: Juan Sola

Known for: Malbec

Signature wine: Luigi Bosca Malbec D.O.C.

Other labels: Finca La Linda, Testimonio, Finca Los Nobles

Visiting: Open to the public by appointment; tasting room

The Arizu family is one of the famous names in Argentinean wine. They have owned the Luigi Bosca winery for more than four generations, and it is one of the most consistent and beloved bodegas in the market. Luigi Bosca was active in the effort to create the Luján de Cuyo DO, and the Malbec D.O.C commemorates that effort. The Arizus own seven vineyards, totaling over 1,700 acres of land spread throughout the region, and they are instigating biodynamic practices in many of their *fincas* (vineyards). The family has also been a leader in the pursuit of quality in the northern province of Catamarca with the Cabernet de los Andes winery.

Manos Negras

Founded: 2009

Address: Prolongación Medrano 2983, Chacras de Coria, Mendoza, www.manosnegras.com.ar

Owners: Jeff Mausbach, Alejandro Sejanovich, Jason Mabbett, Duncan Killiner

Winemakers: Alejandro Sejanovich, Jason Mabbett, Duncan Killiner

Viticulturist: Alejandro Sejanovich

Known for: Malbec, Pinot Noir, Torrontés

Signature wine: Stone Soil Select

Visiting: Open to the public by appointment only

This new and exciting project is a joint venture by three of Argentina's most highly regarded wine professionals. Formerly at Catena, the American Jeff Mausbach drives the marketing and sales. Alejandro Sejanovich, another longtime veteran of the Catena wineries, and the experienced New Zealander Duncan Killiner make the wines. According to Mausbach, the name Manos Negras ("black hands") "pays tribute to the winemakers that roll up their sleeves and get their hands dirty making craft wines. The project arises from the impetus to show Argentina's diversity, with its emblematic varietals and the terroir where we think we could achieve a different expression, with unique characteristics." The team makes an array of wines from vineyards in both Argentina and Chile.

Sejanovich has two other pet projects: in Uco Valley, Zaha, whose name means "heart" in Huarpe, and Anko, in Salta, whose name means "this is dirty work." He works with Mausbach on both projects, as well as Malbec 55 and TintoNegro.

Marcelo Miras

Founded: 2002

Address: Jujuy 2283, General Roca, Río Negro, www.marcelomiras.com.ar

Owners: Sandra Ponce, Marcelo Miras

Winemakers: Marcelo Miras, Pablo Miras

Viticulturist: Marcelo Miras

Known for: Malbec, Pinot Noir, Merlot, Semillon

Signature wine: Miras

Visiting: Open to the public by appointment only

Born into the winemaking trade in Mendoza, Marcelo Miras has been on a mission to promote cooler-climate areas. He was the enologist at Humberto Canale for his first twelve years. With over two decades of experience in Patagonia, he is among the most skilled winemakers in the region and consults for many of the region's wineries. For his own wines, he makes use of selected vineyards in Mainqué and Ingeniero Huergo, in collaboration with his son Pablo. A *garagiste*, he is admired for the elegance and subtlety of his approach.

Masi Tupungato

Founded: 1996

Address: Cap. de Frag. Moyano 53, Mendoza, www.masi.it

Owner: Boscaini family

Winemaker: Masi Technical Group

Known for: Corvina, Malbec, Pinot Grigio, Torrontés

Signature wine: Passo Doble

Visiting: Open to the public by appointment only; tasting room

A fixture in Italy's Veneto, Masi's vineyards have been tended for nearly seven hundred years. In 1964, Masi invented the *ripasso* process (in which lightly fermented Valpolicella is passed over the dried grapes left over from Amarone production) to create Valpolicella Ripasso. This completely new style of wine was awarded its own DOC status. Masi Tupungato, Masi's South American project, makes a wine called "Passo Doble," a blend of Malbec and Corvina, in a smiliar style. In previous years the winery has played with grapes such as Primitivo, Molinara, Rondinella, Nero d'Avola, and Croatina, but it has now focused on marrying a small number of grapes, two Italian varieties (Corvina and Pinot Grigio) and two Argentinean (Malbec and Torrontés).

Melipal

Founded: 2003

Address: Ruta Nacional 7, km 1056, Agrelo, Mendoza, www.bodegamelipal.com

Owner: Aristi family

Winemaker: Victoria Pon

Consulting winemakers: Luis Barraud, Andrea Marchiori

Consulting viticulturist: Andrea Marchiori

Known for: Malbec, Petit Verdot, Cabernet Franc, Cabernet Sauvignon

Signature wine: Melipal Reserva Malbec

Other labels: Ikella

Visiting: Open to the public; tasting room; restaurant

Melipal is based on the Aristi family's acquisition of three vineyards, the oldest dating back to 1923. Though they have no wine background (Ignacio Aristi is a former commodities trader), they went about things methodically and put together an accomplished team. The talented Victoria Pon worked with Paul Hobbs in California. The Hobbs connection is strengthened by the local consulting of Luis Barraud and Andrea Marchiori, Hobbs's partners in Viña Cobos. The winery has a lovely restaurant that focuses on wine and food pairings.

Mendel

Founded: 2004

Address: Terrada 1863, Mayor Drummond, Luján de Cuyo, Mendoza, www.mendel .com.ar

Owners: Anabelle Sielecki, Roberto de la Mota

Winemakers: Roberto de la Mota, Santiago Mayorga

Viticulturist: Santiago Mayorga

Known for: Malbec, Cabernet, Semillon

Signature wine: Unus

Visiting: Open to the public by appointment only; tasting room

This winery is overseen by Roberto de la Mota in partnership with one of Argentina's wealthy diplomats, Anabelle Sielecki. The wines are sourced from eighty-year-old vines in choice vineyards in Perdriel and Mayor Drummond. Discovering and purchasing the Perdriel vineyard was Annabelle's inspiration, and she immediately approached de la Mota to be her winemaker and partner. He has been making wine professionally since he was a teenager, first under the tutelage of his father, later with Weinert, and then with Chandon Argentina, where he helped create Terrazas de los Andes. He began developing Cheval des Andes before leaving to establish Mendel in 2003. Mendel is the first name of Anabelle's father, who came to Argentina with nothing and ended up a successful businessman in different industries. He was a man who loved the finest things life had to offer, particularly wine.

Michel Torino (Bodega El Esteco)

Founded: 1892

Address: Intersection of Ruta Nacional 68 and Ruta Nacional 40, Cafayate, Salta, www.micheltorino.com.ar

Owner: Grupo Peñaflor

Winemaker: Alejandro Pepa

Consulting winemaker: Daniel Pi

Viticulturist: Francisco Tellechea

Known for: Cabernet Sauvignon, Malbec, Tannat, Torrontés

Signature wine: Altimus

Visiting: Open to the public by appointment only; tasting room

The actual winery is called El Etesco, named for a historically significant northern Argentinean city. With over 1,500 acres of vineyards at elevations between 5,500 and 6,600 feet, Michel Torino is the most significant winery in the Cafayate Valley of Argentina. The brothers Salvador and David Michel founded the winery two decades after determining that Cafayate was suitable for fine-wine production. In 2005, the Patios de Cafayate Hotel and Spa was opened. Beyond the expected amenities, it offers interactive harvest programs in the season.

Monteviejo

Founded: 1998

Address: Clodomiro Silva sin número, Vista Flores, Tunuyán, Mendoza, www.monteviejo.com

Owner: Estate of Catherine Péré-Vergé

Winemaker: Marcelo Pelleriti

Consulting winemaker: Michel Rolland

Viticulturists: Carlos Tizio, Marcelo Canatella

Known for: Malbec, Torrontés

Signature wine: Lindaflor La Violeta

Other labels: Mariflor, Festivo

Visiting: Open to the public by appointment only; tasting room; restaurant

This 309-acre Clos de Los Siete project was established by the late Catherine Péré-Vergé, who owned and ran Château Le Gay and Château La Viollette in Pomerol. It is unique in that the winery employs the same technical team, managed by Pelleriti (winemaker) and Michel Rolland, in both Bordeaux and Mendoza. The very affable Marcelo could well be the only Argentinean winemaker in Pomerol. He also makes small amounts of very good wine under his own label (marcelopelleriti.com) and is a lover of Ovation guitars, as his website and wine labels illustrate.

Nieto Senetiner

Founded: 1888

Address: Guardia Vieja sin número, Vistalba, Luján de Cuyo, Mendoza, www.nietosenetiner.com

Owner: Molinos Río de la Plata

Winemaker: Santiago Mayorga

Consulting winemaker: Paul Hobbs

Viticulturist: Tomas Hughes

Known for: Malbec, Bonarda, Cabernet Sauvignon, Pinot Noir

Signature wine: Nieto Terroir Blend

Visiting: Open to the public by appointment only; tasting room; restaurant

Though the winery dates back to 1888, it was purchased in 1969 by Nieto Senetiner, who owned and operated it until 1998. It was then purchased by the Molinos Río de la Plata, one of Argentina's largest food companies. Tomas Hughes works from close to one thousand acres of vineyards and is celebrated for his stellar Bonarda. The winery provides lovely hospitality and a great tour of its historical buildings. You can participate in the harvest (by prior arrangement) and then settle into the traditional Argentinean restaurant. Nieto Senetiner operates Casa Nieto, a small tasting room and café in São Paulo (making it the first Argentinean winery to establish a permanent retail presence in Brazil), and a restaurant, Casa Nieto Buenos Aires, in the chic Recoleta neighborhood of that city.

NQN

Founded: 2001
Address: Ruta Provincial 7, Picada 15,
    San Patricio del Chañar, Neuquén,
    www.bodeganqn.com.ar
Owner: Bodega del Fin del Mundo
Winemakers: Roberto de la Mota, Gustavo
    Agnostini
Known for: Pinot Noir, Malbec, Merlot,
    Cabernet Franc
Signature wine: Malma Universo
Other labels: Malma, Picada 15, Cholila
    Ranch
Visiting: Open to the public; tasting room;
    restaurant and inn (Malma Resto Bar
    and Guest House)

An ultramodern winery, NQN sits on top of a small hill overlooking the vines. The land extends over 3,200 acres, but only a fraction of it is currently being used. With

Fin del Mundo's investment in the winery in 2012, there are plans to plant more vines. The brainchild of former owner Luis María Focaccia and his partner Lucas Nemesio, NQN also runs one of the region's finest restaurants, Malma.

O. Fournier Argentina

Founded: 2002
Address: Los Indios sin número,
    La Consulta, San Carlos, Mendoza,
    www.ofournier.com
Owner: O. Fournier Group
Winemaker: José M. Spisso
Viticulturist: Jorge Nahiem
Known for: Malbec, Tempranillo
Signature wines: Alfa Crux,
    B Crux
Other labels: Urban Uco
Visiting: Open to the public; tasting room;
    restaurant (Urban)

Led by the supremely capable and brilliant José Manuel Ortega Fournier, the O. Fournier Group makes wines in Chile, Spain, and Argentina. A Spanish banker by trade, Fournier first visited Mendoza in 1999, seeking a cool climate, and took a few years to find and acquire his little piece of heaven. Wherever he goes, he makes intense, terroir-driven wines. The winery is an award-winning, 100 percent sustainable architectural gem. A talented and enthusiastic chef, his wife, Nadia, runs Urban, one of the best restaurants in Mendoza, as well as the newer and critically acclaimed Nadia OF restaurant in Chacras de Coria. In an innovative partnership opportunity, the O. Fournier team has offered for sale 85 three-hectare vineyard parcels of a vineyard on an adjoining property. The O. Fournier team does the maintenance, but the investors can produce their own wine and stay in the Fournier guesthouse (at least until plans for a luxury hotel materialize).

Pascual Toso

Founded: 1890
Address: Carril Barrancas sin número, Maipú, Mendoza, www.pascualtoso.com
Owner: Llorente family
Winemaker: Rolando Luppino
Consulting winemaker: Paul Hobbs
Known for: Malbec
Signature wine: Magdalena
Other labels: Finca Pedregal
Visiting: Open to the public by appointment only; tasting room

With Italian roots in Canale d'Alba in Piedmont, Toso has estate vineyards totaling 630 acres and two wineries: a larger one in San José that makes sparkling wines, among other offerings, and a smaller facility in Maipú's Barrancas, where Toso owns some of the best land and focuses on smaller-lot wines. Fourth-generation vintner Enrique Toso is heavily involved in daily operations; he is a gregarious storyteller who is known for hosting the best *asados* in the business for his friends and customers.

Passionate Wine

Founded: 2009
Address: Mitre 79, Tupungato, Mendoza, www.passionatewine.com
Owner: Matías Michelini
Winemaker: Matías Michelini
Viticulturists: Manuel Pelegrina, Matías Michelini
Known for: Bonarda, Pinot Noir, Malbec, Merlot, Cabernet Sauvignon, Sauvignon Blanc, Torrontés
Signature wines: Montesco Parral, Montesco Punta Negra
Other labels: Inéditos, MalBon, Diverso
Visiting: Open to the public by appointment only; tasting room

Matías Michelini is an impressive personality. Starting out with beekeeping at the age of fifteen and moving on to cellar-hand work at seventeen and to winemaking in his later years, this man is indeed passionate. His emphasis on Bonarda is demonstrated in the Parral (a Malbec, Bonarda, and Cabernet Sauvignon blend), MalBon (a Malbec and Bonarda blend), and Inéditos (a pure Bonarda), but the winery also offers a 100% Pinot Noir (Punta Negra) and Sauvignon Blanc (Montesco Agua de Roca). In addition to his work at Passionate, Matías is consulting winemaker for Zorzal (see below).

Patagonian Wines

Founded: 1998
Address: Camino Currumahuida sin número, El Hoyo, Chubut, www.patagonianwines.com
Owner: Bernardo Carlos Weinert
Winemaker: Dario González Maldonado
Viticulturist: Dario González Maldonado

Known for: Merlot, Pinot Noir, Chardonnay, Riesling, Gewürztraminer

Signature wine: Piedra Parada

Visiting: Open to the public; tasting
room

Chubut is perhaps the last region you'd
consider wine country. Home of cattle and
pastureland, it is one of the country's newest wine frontiers. Founded by Bernardo
Carlos Weinert, Patagonian Wines is the
southernmost vineyard in the Americas,
located outside El Bolsón in a small town
called El Hoyo. Its Pinot Noir is rapidly becoming known as one of the country's finest and is definitely worth seeking out.

Pernod Ricard

See Bodegas Etchart, Graffigna

The giant global beverage company Pernod Ricard is a significant player in Argentina, where it has several operational
facilities and three wineries. Each of the
wineries is historically significant and focused on improving the reputation and
quality of the region's wines. Pernod Ricard is a major distributor of Argentina's
wines both domestically and in more than
forty other countries.

Mumm Argentina: In the heart of Mendoza's Uco Valley, Pernod Ricard also produces sparkling wines under the Mumm
Argentina banner. A global offshoot of
the Champagne powerhouse, Mumm
Argentina makes wines mostly by the
Charmat method and intended for daily
enjoyment, though it also produces two
wines using the *méthode traditionnelle:*
Brut Nature (without dosage) and Reserva
Extra Brut.

Piattelli Vineyards

Founded: 2004

Address: Calle Cobos 13710, Agrelo, Luján
de Cuyo, Mendoza, and Ruta Provincial 2 sin número, Cafayate, Salta,
www.piattellivineyards.com

Owner: John Malinski

Winemaker: Valeria Antolín

Consulting winemaker: Roberto de la Mota

Viticulturists: Valeria Antolín, Emilce
Lucentini (Mendoza), Alejandro Nesman
(Salta)

Known for: Malbec, Cabernet Sauvignon,
Torrontés, Chardonnay

Signature wines: Trinità, Premium
Malbec

Other labels: Loscano Vineyards

Visiting: Open to the public; tasting room;
restaurant

With its American owner and talented
winemaking team, advised by Roberto de
la Mota, Piattelli is replete with modern
technology, which stands in stark contrast
to its traditional Mendocino architecture.
Piattelli recently opened a second multi-
million-dollar winery operation in Salta's Cafayate. John Malinski is a wine importer and the owner of a wine-bar group
in Minnesota.

Pulenta Estate

Founded: 1991 (vineyard), 2002 (winery)

Address: Ruta Provincial 86, km 6.5,
Agrelo, Luján de Cuyo, Mendoza,
www.pulentaestate.com

Owners: Hugo Pulenta, Eduardo Pulenta

Winemaker: Javier Lo Forte

Consulting winemaker: Eduardo Pulenta

Viticulturists: Antonio Pulenta, Facundo
   Yazlli (agronomist)
Known for: Malbec, Merlot, Cabernet
   Sauvignon, Cabernet Franc
Signature wine: Pulenta Gran Corte
Other labels: La Flor
Visiting: Open to the public; tasting room;
   restaurant

Pulenta Estate is run by yet another branch
of the famed Pulenta family, the sons of
the viticulturist Antonio Pulenta, who was
the majority shareholder of the Trapiche
winery and is in charge of the vineyards
at Pulenta. Eduardo, the only winemaker
in the third generation of the Pulenta
clan, vinifies his wines from the estate's
336 acres of vines. The old Porsches outside the tasting room are testament to the
family's passion for fast cars. Pulenta Estate has a lovely restaurant and a private
ranch-style house located among the vineyards that is available for small business
gatherings.

Renacer

Founded: 2002
Address: Brandsen 1863, Perdriel, Luján
   de Cuyo, Mendoza, www.bodegarenacer
   .com.ar
Owner: Patricio Reich
Winemaker: Pablo Profili
Consulting winemakers: Alberto Antonini,
   Paul Hobbs
Viticulturists: Marcelo Casazza, Julio
   Correa
Known for: Malbec
Signature wine: Enamore
Other labels: Punto Final

Visiting: Open to the public; tasting room

After retiring from the World Bank, Chilean-born Patricio Reich decided to get
in on the fine-wine boom on the other
side of the Andes. His previous experience as a part owner of Viña San Pedro
fifteen years earlier may explain why he
called his new venture Renacer, meaning
"rebirth." Assisted by the famous consultant Alberto Antonini, Renacer is a bastion of quality. The winery building itself
is impressive, featuring a medieval tower
built of rocks that deftly frames its state-
of-the-art winemaking equipment. Their
Malbec *corte*, Enomare, was inspired by a
visit by Marilisa Allegrini of the famed Ve-
netian winery, who thought that Argentinean fruit lent itself well to using partially dried grapes in the *ripasso* technique
(which traditionally uses Amarone grapes;
the label is a play on the variety name).

Ricardo Santos

Founded: 1992
Address: Maza at Manuel A. Sáenz,
   400 meters north of Ruta Nacional 60,
   Russell, Maipú, Mendoza,
   www.ricardosantos.com
Owners: Ricardo Santos and family
Winemaker: Patricio Santos
Viticulturist: Patricio Santos
Known for: Malbec
Signature wine: El Malbec de Ricardo
   Santos
Visiting: Open to the public by appoint-
   ment only; tasting room

Ricardo Santos is the son of the founder
of Bodega Norton and a former part

owner. Santos started producing wines from a thirty-acre single vineyard—Las Madras, located in Russell, Maipú—exclusively planted to Malbec. "El Malbec de Ricardo Santos," one of his two red wines, is a single-vineyard bottling from this vineyard. He also makes a Semillon and a Cabernet Sauvignon. Ricardo's son Patricio is the winemaker, and his other son, Pedro, directs marketing. Pedro and Patricio Santos are also the proprietors of Tercos, a new winery. The name, which means "stubborn" in Spanish, underscores the brothers' tenacious commitment to quality.

Riglos

Founded: 2002

Address: Ruta Provincial 89, km 13, Tupungato, Mendoza, www.bodegariglos.com

Owners: Dario Werthein, Fabián Suffern, Rafael Calderón

Winemaker: Pulqui Rodríguez Villa

Consulting winemaker: Paul Hobbs

Viticulturist: Hernán Cortegoso

Known for: Malbec, Cabernet Sauvignon, Cabernet Franc

Signature wine: Riglos Gran Corte

Visiting: Open to the public by appointment only

Dario Werthein and Fabián Suffern are childhood friends, but the family connections go back even further: their grandparents grew up together in the small town of Riglos, Argentina. This link is commemorated in the name of their project. Grapes are sourced exclusively from their 178-acre estate vineyard, Finca Las Divas, in Gualtallary. They select about the best 25 per-

cent of the crop for their own use and sell the balance.

Rolland Collection

Founded: 2010

Address: Clodomiro Silva sin número, Vista Flores, Tunuyán, Mendoza, www.rollandcollection.com

Owner: Michel Rolland

Winemakers: Rodolfo Vallebella, Thierry Haberer

Consultant: Michel Rolland

Viticulturist: Marcelo Canatella

Known for: Malbec, Pinot Noir, Sauvignon Blanc

Signature wines: Mariflor, Val de Flores

Visiting: Open to the public by appointment only; tasting room

The Rolland Collection covers a range of operations in different locations. At his Uco Valley winery, adjacent to the Monteviejo, Cuvelier, and DiamAndes wineries, Michel Rolland makes his two Mendocino wines, Mariflor and Val de Flores. Val de Flores is made from grapes from a small, fifty-year-old Malbec vineyard at the foot of the Andes in Vista Flores. Mariflor is Rolland's piece of the Clos de los Siete project, shared with his neighbors.

Rolland's San Pedro de Yacochuya winery is a partnership with Arnaldo Etchart, based in Salta. Casarena, his latest winery project, is majority owned by American investors. Rolland is the adviser, and Gabriela Celeste is in charge of the vineyards and the winery. Its vineyards are located in Tupungato and Agrelo, and the winery is located in Perdriel. The wines were launched in 2011. In addition, Rolland operates Eno Rolland, a viticulture and enology research lab in Mendoza that offers custom-crush, limited-production wines. This too is run by Gabriela Celeste, Michel's right hand in Argentina.

Rutini (Bodega La Rural)

Founded: 1885
Address: Montecaseros 2625, Coquimbito, Maipú, Mendoza, www.rutiniwines.com
Owners: Nicolás Catena, José Benegas Lynch
Winemaker: Mariano di Paola
Consulting winemaker: Paul Hobbs
Viticulturist: Mariano di Paola
Known for: Malbec
Signature wines: Apartado, Encuentro
Other labels: Felipe Rutini, Trumpeter
Visiting: Open to the public; tasting room; museum

Rutini is Argentina's leading restaurant wine brand, sold all over the country. Eighty percent of the wine is sold domestically. Owned by Nicolás Catena in partnership with José Benegas Lynch (a member of the Benegas wine dynasty), Rutini is home to Argentina's most-visited wine museum, La Rural Museo del Vino, which has an astounding collection of wine artifacts, bottles, and other paraphernalia dating back over two hundred years. Rutini has two facilities: the primary one is the older La Rural winery in Maipú, and the other is the new Rutini Tupungato facility. It owns several spectacular vineyards in Maipú, Tupungato, and San Carlos. Mariano di Paola was a professor of enology at the prestigious Don Bosco University before joining the winery in 2007.

Salentein

Founded: 1995
Address: Ruta Provincial 89 sin número, Los Árboles, Tunuyán, Mendoza, www.bodegasalentein.com
Owner: M.P. Wines
Winemaker: José Galante
Consulting winemaker: Paul Hobbs
Viticulturist: Gustavo Soto
Known for: Malbec
Signature wine: Salentein Reserve Malbec
Other labels: Killka, Portillo, Primus, Numina
Visiting: Open to the public; tasting room; restaurant (Killka); inn (Posada Salentein)

This winery is stunning, set in the foothills of the Andes and laid out in the shape of a cross. Salentein is a privately owned estate of almost 5,000 acres, of which 1,124 acres are planted to vine. It has a super restaurant, an impressive art gallery, a splendid visitor center, and a sixteen-room lodge. The Dutch Pon family, owners of M.P. Wines, have created a remarkable little empire that includes Salentein and San Juan's Callia winery.

San Pedro de Yacochuya

Founded: 1999
Address: Ruta Provincial 2, km 6, Cafayate, Salta, www.sanpedrodeyacochuya.com .ar/eng/history.php
Owners: Etchart family, Michel Rolland
Winemakers: Michel Rolland, Marcos Etchart
Known for: Malbec, Cabernet Sauvignon

Signature wine: Yacochuya

Visiting: Open to the public by appointment only; tasting room

Arnaldo Etchart first brought Michel Rolland to his eponymous winery in Cafayate in the late 1980s. They formed a great working relationship, and after the sale of Etchart, Michel and Arnaldo created a partnership to make the wines of San Pedro de Yacochuya. Yacochuya, which means "clear water" in Quechua, is the name of a vineyard in Valles Calchaquíes with sixty-year-old vines.

Terrazas de los Andes

Founded: 1999

Address: Calle Salguero at Ruta Provincial 15, Perdriel, Luján de Cuyo, Mendoza, www.terrazasdelosandes.com

Owner: LVMH

Winemakers: Hervé Birnie-Scott, Nicolas Audebert

Viticulturist: Gustavo Ursomarso

Known for: Malbec, Cabernet Sauvignon, Chardonnay, Torrontés

Signature wine: Terrazas Reserva Malbec

Other labels: Altos de la Plata

Visiting: Open to the public by appointment only; tasting room; restaurant and guesthouse

The original Terrazas property, which has a distinctly Spanish feel, dates back to 1898; it was built by Sorteiro Arizu, who owned almost ten thousand acres of vines here. In the early twentieth century the property was owned by the Spanish sherry and brandy powerhouse Domecq, which built a new winery to accommodate "giraffe" distillation columns for making brandy. The estate's vines were planted in 1930. In the 1990s Terrazas de los Andes was established by Moët Hennessy (now part of LVMH). The grounds house a winery, restaurant, and guesthouse.

Tikal Winery

Founded: 1999

Address: Calle Cobos sin número, Agrelo, Luján de Cuyo, Mendoza, www.tikalwines.com

Owner: Ernesto Catena

Winemaker: Bernardo Bossi

Known for: Malbec, Bonarda

Signature wine: Patriota

Visiting: Open to the public by appointment only

Tikal is named after Ernesto Catena's son and defined by Malbec. Each wine is uniquely sourced or blended, and each is named for a hedonistic or emotional quality. Amorio, for example, means "love affair," and Jubilo means "rejoice." Locura ("madness") is a blend of two red grapes and one white, and Patriota is a "patriotic" blend of Bonarda and Malbec, grapes that have been at the core of Argentine winemaking since its inception.

Trapezio

Founded: 2001

Address: Costa Flores sin número, Agrelo, Luján de Cuyo, Mendoza, www.trapezio .com.ar

Owner: Mauro Villarejo

Winemaker: Leonardo Quercetti

Consulting winemaker: Lucas Mendoza

Viticulturist: Ricardo García

Known for: Malbec, Cabernet Franc

Signature wine: Bo Bó

Other labels: Plus

Visiting: Open to the public by appointment only; tasting room

Trapezio's whimsical website reveals the winery's playful yet focused approach: it has a hand-drawn feel yet shows attention to detail in, for example, its La Promesa vineyard plantings and drip-irrigation practices. The winery has a very cool "adopt a row" program, which entitles renters to receive three to five cases of wine a year for an annual fee related to the production costs. The costs of shipping and customs duties from Mendoza to Madrid or Bordeaux are included in the price. (If you're outside Spain or France, you'll have to inquire about terms.)

Trapiche

Founded: 1883

Address: Calle Nueva Mayorga sin número, Coquimbito, Maipú, Mendoza, www.trapiche.com.ar

Owner: Bemberg family

Winemaker: Daniel Pi

Consulting winemaker: Alberto Antonini

Viticulturist: Marcelo Belmonte

Known for: Malbec, Cabernet Sauvignon, Bonarda, Pinot Noir, Cabernet Franc, Torrontés, Chardonnay

Signature wine: Malbec Single Vineyard Series

Other labels: Broquel, Manos, Iscay, Finca Las Palmas, Zaphy, Astica, Septiembre

Visiting: Open to the public by appointment only; tasting room

Trapiche is the leading export winery of Argentina, selling in over eighty countries. It offers something for every wine consumer—delightful, great-value entry-level wines and several of the leading site-specific wines. In November 2008, the estate celebrated its 125th anniversary with the launch of its new winery building. Daniel Pi is considered one of world's great winemakers. He has a small personal project called 3/14, a line of less than three hundred cases, named for March 14, World Pi Day—an annual celebration of the mathematical constant with which he shares a name.

Trivento

Founded: 1996

Address: Canal Pescara 9347, Russell, Maipú, Mendoza, www.trivento.com

Owner: Concha y Toro

Winemakers: German di Cesare, Victoria Prandina, Rafael Miranda, Fernando Piottante, Maximiliano Ortiz

Viticulturist: Cristian Linares

Consulting winemaker: Alberto Antonini

Known for: Malbec, Cabernet Sauvignon, Syrah, Bonarda, Chardonnay, Torrontés, Viognier

Signature wine: Eolo Malbec

Other labels: Tribu, Brisa de Otoño, Eolo, Colección Fincas, Amado Sur

Visiting: Open to the public; tasting room

In 1990, Concha y Toro, Chile's leading wine producer, announced its acquisition of 3,185 acres in Mendoza. There was little doubt on either side of the Andes that the winds of change were blowing, and it was fitting that the new venture was christened Trivento, a reference to the three winds that are such a distinctive attribute of Mendoza's climate. Supported by the parent company's huge infrastructure, Trivento is a widely distributed and exported brand. Alberto Antonini's deft hand can be tasted in the wines.

Valle de la Puerta

Founded: 1994

Address: Ruta Nacional 74, km 1186, Vichigasta, Chilecito, La Rioja, www.valledelapuerta.com

Owners: International Time Group (ITG) SA, Enrique Liberman

Winemakers: Javier Collovati, Alberto Neira

Consulting winemaker: Mauricio Lorca

Viticulturist: Javier Collovati

Known for: Malbec, Torrontés, Bonarda

Signature wine: La Puerta Gran Reserva

Other labels: Ichanka

Visiting: Open to the public; tasting room

Established as the Valle de la Puerta Olive Grove and Vineyard, this 750-acre property sits between the Famatina and Velasco mountain ranges at more than 3,200 feet above sea level. La Puerta is the intriguing story of three Argentine compadres who share a passion for polo as well as fine wine. Enrique Liberman is the managing principal, who since La Puerta's inception has spared no expense in putting it on the wine world's radar screen. He is joined by the brother of Carlos Menem, ex-president of Argentina, and another friend. They have engaged the assistance of the very talented enologist Mauricio Lorca. Two hundred fifty acres of this sprawling estate are devoted to premium grape varieties; the rest turns out some of Argentina's finest olives, olive oil, and succulent stone fruit, such as peaches and plums.

Viña Alicia

Founded: 1996

Address: Terrada at Anchorena, Mayor Drumond, Luján de Cuyo, Mendoza, vinaalicia.com

Owners: Alicia Arizu, Rodrigo Arizu

Winemaker: Familia Arizu

Viticulturist: Alberto H. Arizu

Known for: Malbec

Signature wine: Colección de Familia Brote Negro

Other labels: Paso de Piedra, Colección de Familia

Visiting: Open to the public by appointment only; tasting room

Alicia Mateu Arizu, wife of the well-known winemaker Alberto Arizu (of Luigi Bosca fame), began producing wines bearing her name after twenty-five years of on-the-job experience. Maintaining a family approach, she undertook this effort with the youngest of her three sons, Rodrigo Arizu. The winery estate is surrounded by a twenty-five-acre vineyard planted with Cabernet Sauvignon, Cabernet Franc, Merlot, and Petit Verdot grapes, along with some vines that are rarely found in Argentina, such as Nebbiolo, Grenache, and Carignan. Viña Alicia makes a few unusual *cortes* that are especially compelling, including "Cuarzo"—a blend of Cabernet Franc, Grenache, and Carignan—as well as a 100 percent Nebbiolo.

Viña Cobos

Founded: 1998

Address: Calle Costa Flores at Ruta Nacional 7, Perdriel, Luján de Cuyo, Mendoza, www.vinacobos.com

Owners: Andrea Marchiori, Luis Barraud, Paul Hobbs

Winemaker: Noelia Torres

Viticulturists: Andrea Marchiori, Valeria Bonomo

Known for: Malbec, Cabernet Sauvignon

Signature wines: Cobos Malbec, Cobos Volturno

Visiting: Open to the public by appointment only; tasting room

This project, Paul Hobbs's baby in South America, is a collaboration with the team of Luis Barraud and Andrea Marchiori, two winemakers with long experience and a history of excellence. The primary 133-acre vineyard, Marchiori, is owned by Andrea's father, Nico, and the Viña Cobos "Bramare" wines (meaning "to yearn" in Italian) are sourced exclusively from this estate. Luis and Andrea have their own small line of wines, Marchiori & Barraud, making about two thousand cases from the Marchiori vineyard, and Luis also consults for Mendoza's Melipal and La Rioja's Chañarmuyo Estate.

Vistaflores Estate

Founded: 2006

Address: Emilio Civit 40, piso 8, Mendoza, www.vistafloresestate.com

Owners: Bill Smith, Jan Soderberg

Winemaker: Attilio Pagli

Viticulturist: Nico Hughes

Known for: Malbec, Cabernet Sauvignon, Cabernet Franc, Syrah

Signature wines: Altamira Navigato family selection, Altamira de los Andes

Other labels: Primmacepa

Visiting: Open to the public by appointment only

The winery name is Vistaflores Estate, but the wines are labeled Altamira. Bill Smith is from Los Angeles, and his family background is not in wine but in equestrianism. Jan Soderberg is a Chicago native and Wall Street veteran. The very capable team benefits from the consulting services of Attilio Pagli, one of Argentina's most respected winemakers. They work from estate-owned vineyard properties in the Uco Valley, between Vista Flores and La Consulta.

XumeK

Founded: 1998

Address: Laprida e Irazoque sin número, Zonda, San Juan, www.xumek.com

Owner: Ezequiel Eskenazi Storey

Winemaker: Daniel Ekkert

Known for: Syrah, Malbec

Signature wine: XumeK

Other labels: Zonda, Granaderos, Viuda Negra

Visiting: Open to the public by appointment only; tasting room

Arriving at XumeK, you are met by a life-size reproduction of a whale, constructed by the artist Adrian Villar Rojas as a tribute to the site. Although it now sits more than 2,625 feet above sea level, it was once a seabed. You'll also find (living) llamas and some miniature ostrich. All of this is the dream of Ezequiel Eskenazi, the approachable, animal-loving, olive-oil-producing owner and founder of XumeK. After selling grapes, he started his own label in 2002 with a specialty in Syrah.

Zorzal Wines

Founded: 2007–8
Address: Camino Estancia Silva sin número, Gualtallary, Tupungato, Mendoza, www.zorzalwines.com
Owners: Canadian investors, Michelini Bros.
Winemaker: Juan Pablo Michelini
Consulting winemaker: Matías Michelini
Viticulturists: Juan Pablo Michelini, Matías Michelini
Known for: Malbec, Pinot Noir, Cabernet Sauvignon, Sauvignon Blanc, Chardonnay, Torrontés
Signature wines: Terroir Único, Gran Terroir, Field Blend
Other labels: Climax
Visiting: Open to the public; tasting room

Zorzal was initially launched by a group of Canadian investors as a fun project. The financial officer, Dan Huras, fell in love with Argentina when he visited, and when an investment banker told him that the Michelini family had purchased 173 acres in a prime grape-growing area and needed money to build a winery, Huras looked at the numbers and jumped at the chance. Using premium grapes from neighbors until their own vines mature, the Michelinis (who also operate Passionate Wines) have been remarkably successful. Zorzal is the name of a local mountain.

4

CHILE

A *huaso* (Chilean cowboy) next to Pinot Noir vines in
Lapostolle's Casablanca vineyard. Photo by Matt Wilson.

Planted acreage: 506,500

World rank in acreage: 10

Number of wineries: 240

Per capita consumption (liters)/world
ranking: 16/35

Leading white grapes: Chardonnay,
Muscat, Sauvignon Blanc, Sauvignon Vert

Leading red grapes: Cabernet Sauvignon, Carmenère, Merlot, Pais,
Syrah

Memorable recent vintages: 2001,
2005, 2007, 2008, 2010, 2011, 2013

INTRODUCTION

A few years ago the wine critic and Master of Wine Tim Atkin described Chilean wine
as the Volvo of the wine industry: dependable, consistent, and mass produced, but not
likely, in his words, to "quicken the pulse." To further his point, he quoted one winemaker as saying, "Chilean wine comes from the checkbook, not the heart." There may be

some truth in these observations, but, to pursue the car metaphor, maybe we should ask whether he has driven a Ford lately. Chilean wine may not yet be a BMW, the ultimate driving machine, but it is making dramatic progress. And Tim himself acknowledged, after a significant June 2012 tasting, that Chile's wine image is changing, "arguably at a faster rate than neighboring Argentina."[*]

Don't let Chile's tranquil persona fool you. In business, Chileans are among the world's most formidable trading partners, be it in copper, vegetables, or wine. Exports account for more than one-third of the gross domestic product. Chile is the world's fifth-largest wine exporter, the second-largest in the New World after Australia (which it may soon surpass), and ships 70 percent of its wine overseas to more than 150 countries.[†] This success is due to the combination of proactive government policies, working free-trade agreements, and international business savvy on the part of the Chilean wine companies. As in Australia, it helps that power lies in the hands of a few. Four out of five Chilean bottles are sold by the three biggest companies: Concha y Toro, Santa Rita, and VSPT.[‡]

These three companies are formidable. They own wineries of all shapes and sizes, in every wine region of Chile. This consolidated approach was very important to Chile in the early 1990s, when offering quality wine at a good price distinguished Chile from many other exporters.

Running nearly three thousand miles along the west coast of South America, from Iquique in the north to Punta Arenas in the south, with an average width of one hundred miles, stick-thin Chile is separated from the remainder of the continent by the Andes. Most of the wine production is located south of the capital city, Santiago, but wine is produced even in the arid north. The cold Humboldt Current in the Pacific Ocean is indisputably the principal climatic influence, cooling the vines from offshore. Even though the Coastal Mountains (between the Central Valley and the Pacific) provide some shelter, the combination of the ocean current and the warmth of the interior valley draws cool sea air inland through the river valleys. To the far north, near Atacama and Coquimbo, it can get blistering hot, and the irrigated vineyards are the primary source of Muscat for the country's extensive production of pisco. In Chilean agriculture, irrigation is essential: rainfall, though higher than in much of neighboring Argentina, is sparse, especially in the foothills on the western slopes of the Andes.

Unlike their Argentinean neighbors, Chileans are not dedicated wine drinkers. Statistics show a lower per capita consumption of wine and a taste for beverages like beer and pisco. There are also eighty rum producers in Chile, and cocktails are gaining in popularity, as a hop through the bars and restaurants of Santiago will attest.

---

[*] Tim Atkin, "Carignan: Chile Returns to Its Roots," *Tim Atkin MW* (blog), www.timatkin.com, July 20, 2012.

[†] Steven Spurrier, "Spurrier's World," *Decanter* magazine, May 2011.

[‡] Kym Anderson and Signe Nelgen, *Global Wine Markets, 1961 to 2009: A Statistical Compendium* (Adelaide: University of Adelaide Press, 2009).

LOCAL HISTORY

Vineyards were first established in Chile in the sixteenth century by the Spanish. Chile's independence from Spain in 1810 and new mining wealth helped shepherd in the so-called French phase of wine culture. Because there were no cars, newly rich Chilean mineral barons were limited in how far they could travel by land from their homes in Santiago. They did, however, visit France; and, noting the status value of French vineyards, they planted their own not far from Santiago in the Central Valley, including Viña La Rosa (established in 1824), Carta Vieja (1825), Santa Rita (1880), Concha y Toro (1883), and Undurraga (1885). Grapes like Pais, planted by the Spanish but now considered inferior, were ripped out and replaced with quality varieties like Cabernet Sauvignon, Merlot, Carmenère, Chardonnay, and Sauvignon Blanc (along with Sauvignon Vert). It was also de rigueur to call in French enologists to assist. In 1830, Claude Gay pioneered a more systematic approach toward cultivating French grape varieties. Nonetheless, several historians consider the most influential wine adviser to be the Chilean-born Silvestre Ochagavía Echazarreta, who opened a winery in 1851. He imported many of the French vines that established the basis of today's trade. His shipments fortuitously predated the arrival of phylloxera in Europe and were "pure ungrafted original" stock. Even today, the absence of phylloxera in Chile eliminates the requirement for grafted vines. To replant, most viticulturists use a system called *mugrón*, taking a cane from an existing vine and directing it back into the ground to develop roots and propagate a new plant, a practice also used in Argentina's flood-irrigated vineyards.

As phylloxera devastated the vineyards of Europe, Chile rose to power by selling wine to thirsty European consumers. This opportunity created a brief golden age for Chilean wine. By the 1930s, Chile's success had got the best of its winemakers, and volume was prized over quality. Politics intervened as well: the government, emboldened by a temperance movement that grew as wine became cheap and abundant, passed a law in 1938 capping production at 60 liters (15.8 gallons) per capita and imposing heavy fines on anyone caught planting vines. This brought the industry to a screeching halt. Amazingly, these laws remained in place until 1974. Chilean winemaking enjoyed a respite during the Second World War, when European exports to the United States stopped, but for most of this period Chilean exports were minimal. The political upheaval and oppression of the 1970s and 1980s did nothing to improve the climate for producing or enjoying wine in Chile.

These were not, however, completely dark decades for the wine industry. The Spanish wine expert Miguel Torres arrived in 1979, while the military dictator Augusto Pinochet was still in power, bringing a European influence that led winemakers to seek out the right terroirs, ameliorate the selection of *vinifera* vines, and import modern equipment. By the end of Pinochet's regime, and with a stable economy emerging, Chilean winemaking was buoyed by international investment and the hope of a steady civilian government.

In 1989, Chilean exports increased by 60 percent over the previous year, with another 50 percent increase in 1990 and a further 50 percent in 1991. Agustín Huneeus Sr., the retired proprietor of the Veramonte and Neyen wineries, told me that "Chile is perhaps the oldest New World country with a fine wine tradition, . . . but it is the newest because really it starts in 1990. What you know of the Chilean industry today is 1990 and beyond." Indeed, the subsequent years were like a fairytale that lasted until the new millennium. Then the Chilean tidal wave ran into increased competition, sagging sales, and global economic challenges. The image of Chilean wine—or really, the lack of one—exacerbates the challenge. Until recently, most global wine drinkers saw Chile as a source of basic, good-value wines and nothing more.

However, I believe the fairytale will have a sequel. Signs of a renaissance in Chilean wine can be tasted in many of the new terroir-focused wines being produced in new regions by gifted young winemakers.

CHILEAN GAME CHANGERS

ÁLVARO ESPINOZA

The organic and biodynamic movements are more prominent in Chile than anywhere else in South America, and the person mainly responsible for that growth is Álvaro

Espinoza. After graduating in agronomy at Santiago's Catholic University, he looked for ways to apply his expertise in organic horticulture to winemaking. He was further influenced by time spent on various trips with Paul Dolan and Jim Fetzer in Northern California between 1993 and 1998. But Álvaro's skills go beyond applying compost teas and invoking the influences of the moon and planets: he was also, presciently, the first to champion and bottle Carmenère as a single variety in 1994. As a young winemaker, he made his way to Bordeaux and returned to spend time at several Chilean estates, including Domaine Oriental and Carmen (where he bottled that first Carmenère), before starting his Geo Wines consulting business, which now works with Emiliana and Leyda, among others. Álvaro is also the proprietor of the critically acclaimed Antiyal winery.

MARÍA LUZ MARÍN

Though I have not had the pleasure of meeting María Luz, the only woman in Chile who is both the founder and the owner of a winery, her leadership is incontestable. In 1999, her visionary acumen brought her into the vanguard of winemakers opening up the newly discovered area of San Antonio. María Luz's belief that a chilly, windblown hillside plot a few miles from the Pacific Ocean would make a promising winery was initially dismissed. But after researching sites in New Zealand, South Africa, and California, she was convinced she could succeed. The hilly slopes, coastal influence, and offshore breezes of San Antonio were magical. She is a true innovator in a country that could do with a few more risk takers, and her visions have been validated by many others who now source grapes from this region.

AURELIO MONTES

It's hard to express in a paragraph the extent of Aurelio's impact on modern Chilean wine. Rightly recognized as Chile's patriarch of quality winemaking, Aurelio studied agronomy and enology at the Catholic University of Chile in Santiago and worked at Undurraga and San Pedro in the early 1970s before realizing that "we couldn't fight against France, Spain, Italy, and so on. So after some years that I tried to push my people towards something better, I decided to quit my job." In 1988, with two business partners, he started Discover Wines, a company that later became Montes Wines. They started with the Montes Alpha line. Some twenty-five years later, Montes has multiple lines of wines from four Chilean districts and is credited with first realizing the potential of Colchagua's Apalta district, one of the best locations for red wines in Chile. Aurelio makes superior-quality wines across the Americas, including Kaiken (in Argentina), Napa Angel (in the Napa Valley), Star Angel (in Paso Robles), and, in Chile, Outer Limits (improbably located in the beach-resort area of Zapallar, north of Aconcagua). In 2013 he was named international vice president of Wines of Chile, in which role he serves as Chile's chief wine ambassador.

IT TAKES A GOOD WOMAN

In 1994, when Alexandra Marnier Lapostolle founded the winery that bears her name, Chile's wine industry was a distinctly macho world. She was (and is) visionary and courageous. Today it is said that Chile leads every country except the United States in the number of women involved at every level of winemaking and marketing: 35 percent of the enologists and enology students in Chile are women. The earliest pioneers—including Cecilia Torres (of Santa Rita and Casa Real), María Luz Marín (of her eponymous Casa Marín), María del Pilar González (a renowned consultant for Viña Chocalán and other wineries), and Adriana Cerda (of Meli)—inspired a generation of other gifted winemakers, including Irene Paiva (I-Latina), Ana María Cumsille (Altaïr), Laurence Real (Las Niñas), and Carolina Arnello (Portal del Alto). And they have all paid it forward by encouraging other talented young women, such as Cecilia Guzmán (Haras de Pirque), Viviana Navarrete (Leyda), Evelyn Vidal (Kingston), Ximena Pacheco (Viña Casablanca), Tamara de Baeremaecker (Concha y Toro), Johana Pereira (Bisquertt and Estampa), Daniela Gillmore (Gillmore), and many more.

PEDRO PARRA

To call Pedro Parra a terroir whisperer is convenient but understates his impact on the Chilean, and indeed South American, understanding of the interaction of soil and vine. Maybe *Wine Spectator*'s calling him a modern-day Indiana Jones is more apt. Having never studied anything related to wine in school, Pedro immersed himself in agricultural science in Montpellier, made his way to Paris, and came back to Chile briefly before returning to France in 2001 to do a doctorate on the impact of soil on wine that included two years of digging and studying pits in Bordeaux and Burgundy. When he returned to Chile, his early work benefited Lapostolle, Montes, De Martino, and Matetic. Known for his ability to identify which varieties will do well in different soils and regions, Parra has also advised wineries in Argentina (Lagarde, Cobos, Altos Las Hormigas, and Zuccardi) while maintaining a few clients in Chile (including William Fèvre Chile, Errázuriz, and Amayna). He is also a partner with François Massoc in Aristos and with some other friends in Clos des Fous.

REGIONS AND WINERIES

Chile's modern focus on small-scale production was rewarded at the 2012 *Decanter* Wine awards, where Chilean wines won five of the twenty-eight international trophies, more

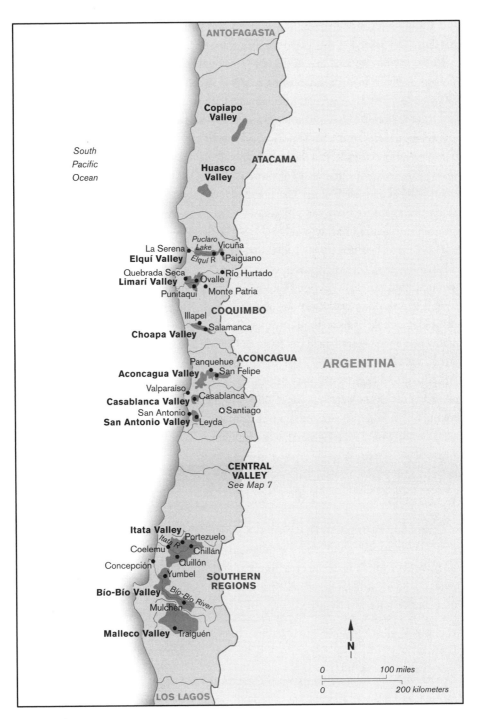

ANTOFAGASTA

Copiapo Valley

South
Pacific
Ocean

ATACAMA

Huasco Valley

Puclaro Lake Vicuña
La Serena
Elquí Valley *Elquí R* Paiguano
Quebrada Seca Río Hurtado
Limarí Valley Ovalle
Punitaqui Monte Patria

COQUIMBO

Illapel
Salamanca
Choapa Valley

Panquehue **ACONCAGUA**
Aconcagua Valley San Felipe

ARGENTINA

Valparaíso
Casablanca Valley Casablanca
San Antonio ✪ Santiago
San Antonio Valley Leyda

CENTRAL VALLEY
See Map 7

Itata Valley Portezuelo
Coelemu *Itata R* Chillán
Concepción Quillón
Yumbel
SOUTHERN REGIONS
Bío-Bío Valley *Bío-Bío River*
Mulchén

Malleco Valley Traiguén

N

0   100 miles

0   200 kilometers

LOS LAGOS

Map 6. Chile's wine regions

than any other country. To quote Steven Spurrier, "There doesn't seem to be a single grape that does not succeed in this viticultural paradise, where creativity knows no bounds."*

Chilean wine regions are under the jurisdiction of SAG, the Servicio Agrícola y Ganadero, which in 1994 established a decree setting forth the zoning and standards for *denominacións de origen* (DOs) and dividing the country into five major wine regions, which are further divided into subregions, zones, and areas. With arid heat in the north and icebergs to the south, it's logical to assume that differences in wine styles would be driven mostly by latitude. But the truth is that Chilean soils vary more dramatically from west to east, as the result of tectonic movements, colluvial and alluvial deposits, and soil evolution that date back eons. The noted viticulturist Eduardo Jordan of De Martino claims that there are more soil differences between the Andes and the ocean than there are from northern Limarí to southern Bío-Bío. In September 2012, in belated recognition of these differences, SAG added three new broad-based, complementary appellations: *costa* (coastal areas), *Andes* (mountain areas), and *entre cordilleras* (between Chile's two mountain ranges). Though there are pundits on both sides debating the value of these new, voluntary classifications, they codify the thinking that a lateral rather than latitudinal approach is worthy of consideration. SAG also regulates labeling according to quality and style: wines labeled as Reserva Privada and Gran Reserva must spend time in oak, and all Reserva wines have higher minimum alcohol levels.

Although I recognize the potential merit of the three new appellations, for practical reasons I present the wine regions of Chile in a traditional north-to-south order. Readers looking for an in-depth education on Chilean wine should get hold of anything written by Master of Wine Peter Richards, who could forget more about the country than most of us will ever learn.

ATACAMA

In Chile's far north is the DO of Atacama, which takes its name from the adjacent desert. There are two subregions for grape growing, the Copiapó Valley and the Huasco Valley, both of which focus on criolla grapes and Muscat for pisco production. My sommelier friend Ricardo Grellet notes that there's a promising experimental project in San Pedro de Atacama, so stay tuned.

COQUIMBO

The DO of Coquimbo contains several exciting wine regions. Long thought of, like Atacama, as a bastion of grapes destined only for distillation or the table, Coquimbo is now home to dynamic subregions whose potential has recently been rediscovered. In

---

* "Finding El Dorado," *Decanter,* August 2011.

addition to growing grapes, Coquimbo is redefining itself as a source of high-quality fruit, especially papayas, and other gastronomic specialties, including goat cheese.

The Elquí Valley sits in an arid portion of Chile called the Norte Chico and contains two viticultural areas close to the Andes, Vicuña and Paiguano. Close to the equator, the region lies at the country's narrowest point, with minimal shelter from the Coastal Mountains, which do not rise to significant heights here. It draws scientists as well as winemakers: because of the thin, dry, and clean air, several international observatories are located here. Some New Age visitors are drawn by the magnetic vibes: measurements of Earth's magnetic forces taken from space in 1982 found that they were strongest near the Elquí Valley.[*] This arid subregion relies on irrigation from snowmelt and Puclaro Lake. Cabernet Sauvignon and Syrah are the most widely planted varieties. Peppery Syrah is the region's incontestable star, resembling wines from the northern Rhône, while Elquí's other leading *vinifera* wines include lovely Sauvignon Blanc that is restrained and elegant, with pronounced bay-leaf notes. Elquí has vineyards, such as those near Alcohuas, that extend as high as 6,400 feet, making it home to some of Chile's highest-altitude winemaking.

The Limarí Valley's landscapes and climate are also hot and dry. The Limarí River runs through the heart of the region, bringing snowmelt from the Andean peaks to the region's towns and fields. The Limarí is regulated by a dam just northwest of Ovalle, which supplies water to the vineyards throughout the growing season. The clay and limestone soils found here are rare in Chile: they are the result of former seabed layers raised by tectonic activity into the Andes and gradually washed back downstream by glaciers and rivers to the plains and valleys below. Less than one-fifth of Limarí's grapes go to making quality wines. Chardonnay is the defining grape, especially fruit from the celebrated area of Quebrada Seca, an elevated section of Ovalle on the northern bank of the Limarí, which is capable of producing age-worthy styles of Chardonnay with a signature chalky or saline minerality. Syrah is also successful, with *garrigue*-scented styles coming from the cooler, coastal vineyards, and fuller, more fruit-driven styles emanating from the warmer sites in the east. Limarí also has a few key areas in the Andean districts of Monte Patria and Río Hurtado and the more central Punitaqui. The region is home to some of Chile's best beach towns, shellfish (especially scallops), and astronomy observation posts. It is the birthplace of Chilean pisco.

Coquimbo's third subregion, the Choapa Valley, is located in a thin strip to the south, where the Andes and Coastal Mountains converge. Grapes are grown in Salamanca and Illapel. Although there are currently no wineries in the area, the vineyards, planted on very rocky soils, are producing peppery Syrah and concentrated Cabernet Sauvignon. Definitely a location to keep in mind, Choapa promises big things from its tiny 330 acres.

---

[*] Jade Frank, "Chile's Elqui Valley: A Zen Experience in 'Travel Therapy,'" *GoNomad*, n.d., www.gonomad.com/3006-chile-s-elqui-valley-a-zen-experience-in-travel-therapy#ixzz2qbNZa3M1, accessed January 17, 2014.

Dalbosco   OchoTierras

Domaine de Manson   Tabalí

Falernia   Tamaya

Maycas del Limarí (*see also* Concha y
   Toro)

ACONCAGUA

Situated sixty miles north of Santiago, the Aconcagua DO takes its name from the Acon-
cagua River rather than the 22,800-foot peak in Argentina. From north to south, the
DO's subregions include Casablanca and San Antonio on the coast, which are among
Chile's coolest new winemaking zones, and the greater Aconcagua Valley inland.

Much of the Aconcagua Valley is considered too hot for grape growing, and Don
Maximiano Errázuriz was laughed at in the mid-1800s when he planted his first vines
in the district of Panquehue. He had the last laugh: Viña Errázuriz's "Seña," an iconic
Bordeaux-style blend that is considered one of Chile's best, is still made there. Greater
Aconcagua, especially the so-called Aconcagua Costa area, is in transition, seeking to
shed its classic but dull image: however, the two coastal subregions described below give
this DO a reputation for quality.

RECOMMENDED PRODUCERS

Arboleda   Seña

Caliterra   Viña San Esteban

Errázuriz   Viña von Siebenthal

Flaherty   Viñedo Chadwick

Casablanca Valley

Casablanca, Chile's first cool-climate coastal region, was initially planted to vine in the
early 1980s. Pablo Morandé, formerly winemaker at Concha y Toro, noted the potential
of this area of pastureland (where he grew up) while seeking out cooler-climate opportunities. Because the area is dry, Concha y Toro backed off, but Pablo was committed,
and he left the company in the mid-1990s in a bold move to develop the region. Limited
irrigation water constrains the region's ability to expand: there is no river, and water
must be pumped from artesian wells. Even so, Pablo's success attracted new investors
and began the quest for expanded coastal plantings.

No more than twenty miles from the Pacific Ocean at any point, Casablanca is strongly influenced by the effects of the Humboldt Current. Cooling breezes blow in from the ocean toward the mountains, filling the vacuum created by warm air rising in the east. The reverse winds that blow in the evening, as colder air flows down the mountain slopes, are not strong enough to provide a cool conclusion to Casablanca's days. (A similar maritime influence brings morning fogs in California's Russian River Valley, especially in the winter.) The resulting climate leads to a long ripening season. The additional time on the vine allows the grapes to develop flavor complexity while maintaining excellent acidity.

The planted area of fourteen thousand acres is about 65 percent white (Sauvignon Blanc and Chardonnay), with the balance being mostly Pinot Noir, Merlot, and Syrah. The last is the crown jewel of the western vineyards. Though unofficial, subareas of Casablanca include Las Dichas (the coolest location, with harvests two weeks later than the rest of the region), La Vinilla (a little warmer), Alto and Bajo Tapihue (similar to La Vinilla), and Lo Ovalle (the warmest). Only 25 percent of the total area is suited for grape cultivation. A dozen or so wineries in the region use about 60 percent of the fruit; the rest is bought by wineries elsewhere. Casablanca Sauvignon Blanc, the region's signature wine, has one style in the cooler western regions (Bajo Casablanca) that is linear, citric, a bit New Zealand and herbal and a little bit Santa Ynez (with notes of green olive and light elderflower), and another in the warmer areas of the east (Alto Casablanca), where it is usually redolent of tangerine, ripe melon, and papaya. Chardonnay is mostly limited to the eastern valley. Casablanca's cool-climate Syrah is intensely stylized, but if you like that spicy style, it is very impressive. Pinot Noir, the acclaimed regional red, has cherry and red-raspberry notes with a linear penetration of acidity akin to the Sauvignon Blanc.

RECOMMENDED PRODUCERS

Casas del Bosque

Catrala

Emiliana

Kingston Family Vineyards

Loma Larga

Morandé

Quintay

Veramonte

Villard

Viña Casablanca (*see also* Carolina
   Wine Brands)

Viña Mar (*see also* VSPT
   Wine Group)

William Cole Vineyards

San Antonio Valley

West of Santiago and immediately south of Casablanca is the small coastal wine region of the San Antonio Valley. A very new addition to Chilean viticulture (with commercial plantings established only in the late 1990s), the region distinguishes itself, like Casablanca, with its ability to grow quality Pinot Noir in addition to its critically acclaimed

Sauvignon Blanc and Chardonnay. The climate of the San Antonio Valley is also profoundly influenced by the Pacific Ocean, with cool morning fog shepherded in by the Humboldt Current. With 5,800 planted acres, this location is home to a small number of individual producers rather than wineries focused on mass production. The valley has a few subdistricts, but the principal zone is that of the Leyda Valley, which was the first trial area, planted in 1988. Just fifty-five miles west of Santiago, the Leyda Valley produces stellar cooler-climate wine styles and has been granted the status of an independent appellation within the San Antonio Valley subregion. The landscape consists mostly of rolling hills, with an average elevation of 600 feet, with clay loam soils over a granite base. Known for a richer and oilier style than the more citric versions in Casablanca, Leyda's Sauvignon Blancs are fleshy, with notes of fresh basil and sorrel and a pronounced earthiness. The unofficial Lo Abarca area of Leyda is acknowledged for a Sauvignon Blanc style that is more feral and ripe papaya: it's love-it-or-hate-it stuff, with subtle tones of honey and fennel. There are only two producers here (Casa Marín and Undurraga) and one independent grower. Two other Leyda Valley areas worth noting for the future are San Juan and the Marga Marga Valley. The larger communes of Rosario and Malvilla round out the San Antonio Valley, where about half the production is in Sauvignon Blanc and the rest split mostly between Chardonnay and Pinot Noir, with small plantings of Syrah.

RECOMMENDED PRODUCERS

Casa Marín	Ventolera
Leyda (*see also* VSPT Wine Group)	Viña Garcés Silva
Matetic	

CENTRAL VALLEY

Chile's wine scene is dominated by the Central Valley. Accounting for 83 percent of Chile's production, it is segmented into four subvalleys—Maipo (with 11 percent of the national total), Rapel (33 percent), Curicó (15 percent), and Maule (24 percent)—and is located between the Andean foothills to the east and the Coastal Mountains to the west. It is Chile's oldest winemaking region and the epicenter of the wine industry, extending over some 250 miles from north to south. It sits directly across the Andes from Argentina's Mendoza region.

Maipo, a region encompassing almost thirty-one thousand acres of vines (60 percent red), is where it all began. Because of its proximity to Santiago, the Santiaguinos planted this area early, and it remains home to the many of the larger players, including Concha y Toro (Chile's largest producer), Santa Rita, and Cousiño Macul. Maipo

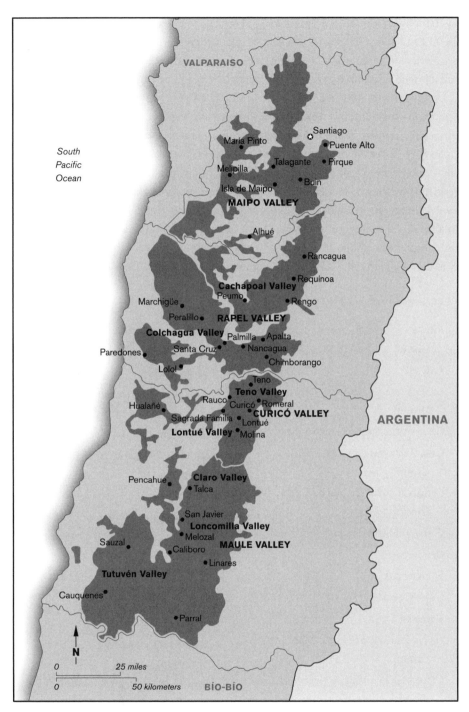

VALPARAISO

South
Pacific
Ocean

María Pinto

Santiago

Puente Alto

Talagante

Pirque

Melipilla

Isla de Maipo

Buin

MAIPO VALLEY

Alhué

Rancagua

Requínoa

Cachapoal Valley

Peumo

Rengo

Marchigüe

Peralillo

RAPEL VALLEY

Colchagua Valley

Palmilla

Apalta

Paredones

Santa Cruz

Nancagua

Chimborango

Lolol

Teno

Teno Valley

Rauco

Curicó

Romeral

Hualañé

CURICÓ VALLEY

Sagrada Familia

Lontué

Lontué Valley

Molina

Pencahue

Claro Valley

Talca

San Javier

Loncomilla Valley

Melozal

MAULE VALLEY

Sauzal

Caliboro

Linares

Tutuvén Valley

Cauquenes

Parral

ARGENTINA

N

| 0 | 25 miles |
| 0 | 50 kilometers |

BÍO-BÍO

Map 7. Chile: The Central Valley

has benefited from an emerging understanding of the area's diversity, especially in the Andean sector known as the Alto Maipo, which sits at an elevation of more than two thousand feet. Much of the country's best Cabernet Sauvignon is grown here, and it is also the source of several of Chile's best red wines, including Almaviva, Concha y Toro's "Don Melchor," and Viñedo Chadwick. The most celebrated areas of the Alto Maipo include Buin, Pirque, and Puente Alto, the star. Located on the southern fringes of Santiago, Puente Alto ("high bridge") is situated at an altitude of roughly 2,300 feet. A nearby bridge over the Maipo River gives the area its name. Almaviva and Don Melchor come from Puente Alto. Alto Maipo produces great Cabernet Sauvignon, reminiscent of Pauillac wines—well structured, with herbal and graphite notes. Central Maipo includes Isla de Maipo, Melipilla, Alhué, María Pinto, and Talagante, a small area that is home to several recognized names, including De Martino, Tarapaca, and Undurraga. Maipo really can do it all: its wines range from super-premium to everyday, white to red, Cabernet Sauvignon to Carmenère to Chardonnay.

RECOMMENDED PRODUCERS

Almaviva (*see also* Concha y Toro Group)

Antiyal

Canepa (*see also* Concha y Toro Group)

Carmen (*see* Santa Rita Group)

Casa Rivas (*see also* VSPT Wine Group)

Cono Sur (*see also* Concha y Toro Group)

Cousiño Macul

De Martino

El Principal

GEO wines

Haras de Pirque

Nativa Eco Wines (*see also* Santa Rita Group)

Ochagavía (*see also* Carolina Wine Brands)

Odfjell

Pérez Cruz

Quebrada de Macúl

Rukumilla

Santa Alicia

Santa Carolina (*see also* Carolina Wine Brands)

Santa Ema

Santa Rita (*see also* Santa Rita Group)

TerraMater

Tres Palacios

Undurraga

Ventisquero

Viña Aquitania

Viña Chocalán

Viña Maipo (*see also* Concha y Toro Group)

Viña Tarapacá (*see also* VSPT Wine Group)

Viñedo Chadwick

William Fèvre Chile

Rapel Valley

The larger Rapel Valley is better known for its individual zones, Cachapoal and Colchagua, than as a single region. Located to the south of Santiago, it is named for the Rapel River, a confluence of the Tinguiririca and the Cachapoal rivers. The latter divides the valley into its two subregions: Colchagua Valley in the south and Cachapoal Valley in the north. Rapel Valley wines are produced predominantly from red varieties, as the dry, warm climate favors Cabernet Sauvignon and Carmenère. Malbec plantings, especially in Colchagua, are on the increase, presumably in an attempt to share in the success enjoyed by Mendoza across the Andes.

Cachapoal   Located between the more prestigious Maipo and Colchagua valleys, Cachapoal (except for Peumo) is to a certain degree neglected; however, many Rapel Valley wines are blends of fruit from both Colchagua and Cachapoal. It has four main areas: Peumo (celebrated for its outstanding Carmenère), Rancagua, Rengo, and Requínoa. Cachapoal is unabashed red-wine territory—some 80 percent of the valley is planted to red varieties—and Merlot and Cabernet Sauvignon perform admirably. It is also a center for fruit and vegetable production, rodeos, and mining. Although Cachapoal lacks the buzz of neighboring Colchagua, its 24,200 acres are nevertheless capable of making very good wines, such as Concha y Toro's iconic Carmenère "Carmín de Peumo." The eastern edge of the wide valley, in the Andean foothills, is home to the noteworthy Alto Cachapoal wineries and vineyards. To the north and northwest are the foothills of the Coastal Mountains, which at this point are closer to the Andes than they are to the coast. The soils of Cachapoal are a mixture of sand, clay, and slightly more fertile loam, which some of the region's winemakers say is ideal for growing Carmenère and Cabernet Sauvignon.

RECOMMENDED PRODUCERS

Altaïr (*see also* VSPT Wine Group)	San José de Apalta
Casas del Toqui	Terraustral
Clos des Fous	Torreón de Paredes
Los Boldos	Vik
Misiones de Rengo (*see also* VSPT Wine Group)	Viña La Rosa
	Viña Tipaume

Colchagua   Colchagua is where the excitement lies. Since its recognition in the late 1990s, the southernmost zone in the Rapel Valley has distinguished itself with its stellar red wines (though it also produces quality fruit for some exciting whites from the new coastal appellation, especially in Paredones). Its seventy-one thousand planted acres make it one of the two largest regions, along with Maule, and it comprises a patchwork of areas, including Peralillo, Nancagua, Chimbarongo, Palmilla, San Fernando,

Marchigüe, Santa Cruz (where you find the renowned but unofficial Apalta), and Lolol, where Paredones is located. Colchagua is slightly cooler than the greater Maipo region to the north, but it maintains a consistently Mediterranean climate. The soil is based on clay and silt loams at the lower elevations, with granitic veins in the higher reaches. While Cabernet Sauvignon is the lead grape, followed by Merlot, the Carmenère is, in my opinion, consistently better here than anywhere else in Chile. It accounts for almost half the country's plantings of this variety. Marcela Chandia, former winemaker at Estampa, notes that Carmenère in Colchagua is "all about blackberry and blueberry fruit, some spice and ripe tannins, and not the bell pepper and green bitterness you find in cooler climates elsewhere." The local Syrah is also excellent, exhibiting a range of styles from ripe raspberry and herbal to meaty and savory.

Apalta is home to two of Chile's most renowned wines: Montes's "Alpha M" and Lapostolle's "Clos Apalta." The other star of the zone is the western Marchigüe area, gently sloping, with that granitic soil that often grows the Carmenère and Cabernet Sauvignon that make Reserva and Gran Reserva wines. I find Carmenère from Colchagua to be the country's best, led by Apalta and Marchigüe, followed by Cachapoal's Peumo and Maipo's Puente Alto and Isla de Maipo.

RECOMMENDED PRODUCERS

Caliterra

Canepa (see also Concha y Toro
    Group)

Casa Silva

Cono Sur (see also Concha y Toro
    Group)

Estampa

Hacienda Araucano

Kankura

Koyle

La Playa

Lapostolle

Laura Hartwig

Los Vascos

Luis Felipe Edwards

Maquis

MontGras (see also MontGras
    Properties)

Neyen

Ninquén (see also MontGras
    Properties)

Polkura

Santa Helena (see also VSPT Wine
    Group)

Viña Montes

Viña Siegel

Viña Tamm

Viñedos Marchigüe

Viu Manent

Curicó Valley

Curicó Valley wins the prize for the sheer number of *vinifera* varieties planted: more than thirty have been grown there since the mid-1800s. It is also the site of Chile's winemaking rebirth in the 1970s, where Miguel Torres began kicking open the doors to international awareness and foreign investment. Interestingly, you don't hear much about Curicó these days, positive or negative: it seems but an afterthought in the latest wave of winemaking excitement. It is divided into two sections, eastern and western. Both are roughly 115 miles south of Santiago. Curicó's climate is divided as clearly as its boundaries; its eastern portion is cooler, benefiting from breezes coming down from the slopes of the Andes. This is in contrast to northern Chile, where the western ends of the valleys are generally cooler because of the effects of the Pacific Ocean. In Curicó, the hills of the Coastal Mountains dissipate the effect of westerly air movement, leaving it warmer than in the east. This may be why the major centers of production and the established names in Curicó Valley winemaking (Echeverría, Montes, San Pedro, Torres, and Valdivieso) are located around the eastern towns of Curicó and Molina.

A source of many old Cabernet Sauvignon vines, the Curicó Valley has ample rainfall (by Chilean standards), well-drained soils, and long, dry summers (surely ideal for many varieties). Curicó has two subzones, which each have a few distinct communes: Teno Valley (Rauco, Hualañé, and Romeral) and Lontué Valley (Molina and Sagrada Familia). Lontué is in the cooler southeast, where the namesake river flows down from the nearby Andes and neatly divides Curicó in two, the area to the north being the Teno region. White-wine production is heavily based on the unintriguing and widely interplanted Sauvignon Vert. The main red variety is Cabernet Sauvignon, which can be very good, especially in Molina.

RECOMMENDED PRODUCERS

Altacima

Altaïr (*see also* VSPT Wine Group)

Aresti

Casa Rivas (*see also* VSPT Wine Group)

Echeverría

Miguel Torres

Misiones de Rengo (*see also* VSPT Wine Group)

San Pedro (*see also* VSPT Wine Group)

Santa Helena (*see also* VSPT Wine Group)

Valdivieso

Viña Requingua

Viña Tarapacá (*see also* VSPT Wine Group)

Maule Valley

Everyone I spoke to about the Maule Valley was effusive in their praise for its potential. Historically, Maule has been viewed as the engine powering Chile's vast production of ordinary wines. With no maritime influence and warm, dry conditions, the region is perfect for growing grapes, and growing a lot of them. For many years winemaking here was focused on squeezing out as much volume as possible. Maule has seventy-nine thousand acres under vine and extends over three zones: the Claro Valley, the Loncomilla Valley, and the Tutuvén Valley. Although Cabernet Sauvignon is the region's most widely cultivated grape, País-based bulk wines intended for local consumption still make up a massive share of the Maule Valley's production.

This vast appellation is split into north and south by the Maule River, which runs west from the Laguna del Maule near the Argentinean border to the Pacific Ocean, and many of the region's vineyards lie adjacent to the river. A new interest in the old vines, especially Carignan in the west, is driving a quixotic revival of this region, led by the members of quality winemaking associations like Vigno and MOVI. The epicenter for this activity is Cauquenes, an area in the Tutuvén Valley. The region is also capable of growing complex, berry-scented Cabernet Sauvignon and very juicy and inky Malbec. Specific areas worth keeping in mind include Loncomilla's Linares and Parral, and Melozal and Sauzal in Tutuvén. Maule is poised for greatness and should soon take its place among the red-wine stars like Colchagua and Alto Maipo.

Botalcura

Calina

Cremaschi Furlotti

Garage Wine Company

Gillmore

J. Bouchon

Laberinto

Meli

O. Fournier Chile

TerraNoble

Via Wines

Viña El Aromo

Vinos del Sur

THE SOUTHERN REGIONS

Chile's Southern Regions DO contains three official DO valleys: Itata, Bío-Bío, and the southernmost, Malleco. Cultivation in these small areas is mostly Pais and Muscat of Alexandria.

The Itata Valley was historically lumped in with the Bío-Bío Valley but is now recognized as its own subregion. Noteworthy for its red soils and copious precipitation (44 inches annually), Itata has significant thermal amplitude. Growers there are slowly moving away from the lower-end Pais and Muscat in favor of Cabernet Sauvignon and Chardonnay. Extending over sixty miles north to south, Itata has four distinct areas: Chillán, Quillón, Portezuelo, and Coelemu. The region is exposed to the elements because the benefits of the Coastal Mountains dissipate toward Concepción. Weather patterns arrive unimpeded from the Pacific Ocean, bringing high rainfall.

In the eyes of most wine experts, Bío-Bío marks the real transition to the deep south. Keep on the path, and your next stop is icebergs and penguins! The favored varieties of the four thousand acres of Bío-Bío (and adjacent Malleco) are classic grapes from Burgundy, Chardonnay and Pinot Noir. This area has been late in developing a fine-wine culture because the rain scared producers away, but now it is becoming well regarded for wines that possess a taut linear essence, akin to those of Limarí (Chardonnay) and San Antonio (Pinot Noir), with distinct minerality. Bío-Bío has also proved to be an excellent spot for aromatic white varieties, notably Riesling, in the vineyards near Mulchén. The cool climate and extended growing season are well suited for developing the signature aromatics of this variety, more so than the hot, dry climates of the Maipo or Rapel. Bío-Bío has two areas for wine in Yumbel and Mulchén.

Finally, the Malleco Valley is the southernmost wine-growing subregion and has extended Chile's wine-growing reach to 40° south, matching the latitude of Argentina's Río Negro. The Malleco Valley is a young but exciting location, and, like Bío-Bío, it seems to be excelling in Chardonnay and Pinot Noir. The one official wine area in Malleco is Traiguén, 385 miles south of Santiago. The richness of Traiguén's soils would normally lead to excessive vigor, but the winds and rain keep fruit yields naturally low.

Fungal diseases and rot can be troublesome, but they are kept in check by the winds' drying effects. The wines here are zippy, zesty, crunchy, and racy, making the wines of Casablanca and San Antonio wine seem almost soft in comparison.

RECOMMENDED PRODUCERS

Gracia de Chile (Agustinos and Veranda labels)

Llai Llai (*see also* Dos Andes)

Porta (*see also* Dos Andes)

WINERY PROFILES

Man plowing between century-old Carignan vines for Garage Wine Company in the Maule region. Photo by Matt Wilson.

Almaviva

Founded: 1997

Address: Av. Santa Rosa 821, Puente Alto, Maipo, www.almavivawinery.com

Owners: Baron Philippe de Rothschild and Concha y Toro Group

Winemaker: Michel Friou

Known for: Cabernet Sauvignon, Carmenère, Cabernet Franc

Signature wine: Almaviva

Visiting: Open to the public by appointment only; tasting room

An Alto Maipo joint venture with the celebrated Baron Philippe de Rothschild company (owners of Château Mouton Rothschild), Almaviva brings you as close to Pauillac as you can get in Chile with its singular red wine, its Bordelaise *chai* (above-ground barrel shed), and its very Old World approach. The name Almaviva, though it sounds Hispanic, is taken from a comic figure in French literature: Count Almaviva in *The Marriage of Figaro* by Beaumarchais, later adapted by Mozart into an opera. The wine is produced under the joint technical supervision of both partners.

AltaCima

Founded: 2001

Address: Longitudinal Sur, km 202, cruce Santa Rosa, Curicó, www.altacima.cl

Owner: Klaus Schröder

Winemaker: Klaus Schröder

Consulting winemaker: Rafael Tirado

Viticulturist: Gonzalo Leal, Víctor Gac

Known for: Cabernet Sauvignon, Carmenère, Syrah, Chardonnay, Gewürztraminer

Signature wine: AC 6330 Ensamblaje

Visiting: Open to the public; tasting room; restaurant

After studying enology and vineyard management in Weinsberg and Geisenheim in Germany, Klaus Schröder arrived in Chile in 1965 as the winemaker for Viña San Pedro. In 1971, he and his wife bought and planted 153 acres of vineyard in the Lontué Valley. AltaCima, which means "high summit," offers a range of personalized tours and a restaurant offering wine and food pairings.

Altaïr

Founded: 2001

Address: Fundo Totihue sin número, Pimpinela-Requínoa, Cachapoal, www.altairwines.com

Owner: VSPT Wine Group

Winemaker: Ana María Cumsille

Consulting winemaker: Pascal Chatonnet

Viticulturist: René Vasquez

Known for: Cabernet Sauvignon, Carmenère, Syrah, Cabernet Franc, Petit Verdot

Signature wine: Sideral

Visiting: Open to the public; tasting room

Altaïr, located roughly sixty miles south of Santiago in Alto Cachapoal, aspires to produce some of the most sought-after, critically acclaimed wines in Chile. The winery was initially a joint venture between Laurent Dassault, owner of Château Dassault and Château La Fleur in Saint-Émilion, Bordeaux, and San Pedro, but Dassault later sold his share in the company back to the San Pedro group. The enologist Ana María Cumsille has global experience, with a résumé that includes stints at Bordeaux's Château Margaux and Château La Louvière. The winery is named for an extremely bright star in the constellation Aquila (the Eagle) that is visible for part of the year in the southern hemisphere.

Antiyal

Founded: 1996

Address: Camino Padre Hurtado, Sitio 68, El Tránsito, Paine, Maipo, www.antiyal.com

Owner: Álvaro Espinoza

Winemaker: Álvaro Espinoza

Viticulturist: Álvaro Espinoza

Known for: Carmenère, Cabernet Sauvignon, Syrah

Signature wine: Antiyal

Other labels: Kuyen

Visiting: Open to the public by appointment only; inn (La Viñita)

This mini-boutique winery is the project of the esteemed Álvaro Espinoza and his family. The name means "sons of the sun" in Mapuche. When he was winemaker at Carmen (see below), Espinoza began making wine at home as a spare-time project, with a single red wine made from Cabernet Sauvignon grapes planted around the family home near Alto Jahuel, as well as some Carmenère and Syrah from his parents' property in Isla de Maipo. A tiny winery was installed next to the house. Although the project has grown, it remains faithful to Espinoza's original goal of small-scale production made from grapes grown according to organic and biodynamic principles.

Arboleda

Founded: 1999
Address: Av. Nueva Tajamar 481, Torre Sur, Oficina 503, Las Condes, Santiago, www.arboledawines.com
Owner: Eduardo Chadwick
Winemaker: Carolina Herrera
Viticulturist: Jorge Figueroa
Known for: Sauvignon Blanc
Signature wine: Arboleda Sauvignon Blanc
Visiting: Open to the public by appointment only; tasting room

Arboleda is Eduardo Chadwick's personal project in the Valle de Aconcagua, created as a tribute to Chile's native trees, which have been preserved in his sustainably managed vineyards. It was originally a part of a Mondavi joint venture, along with Caliterra in Colchagua, and offers a range of boutique wines from specific terroirs.

Aresti

Founded: 1951
Address: Bellavista sin número, Río Claro, Molina, Curicó, www.arestichile.cl
Owner: Ana María y Begoña Aresti López
Winemaker: Jon Usabiaga
Viticulturist: Roberto Turedo
Known for: Syrah, Gewürztraminer, Sauvignon Blanc, Cabernet Sauvignon
Signature wine: Aresti Family Collection
Other labels: Trisquel, Art

Visiting: Open to the public by appointment only

Though established in Curicó as growers some forty years earlier, in 1999 the Aresti family began making wine under their own label, acquiring new land and investing in winemaking technology. Today, three generations of the Aresti family participate. The Family Collection line is adorned with Alberto Valenzuela Llanos's feted 1908 painting *Harvest at the Bellavista Hacienda,* which won a gold medal in an exhibition in the Bellas Artes Museum in Santiago during Chile's centennial celebration. The Arestis acquired the painting and turned it into the symbol of the winery. Aresti is certified to the international environmental-management standard ISO 4001 and IMO organically accredited for specific vineyard sites on its one thousand cultivated acres.

Bisquertt

Founded: 1980
Address: Av. El Condor Sur 590, Oficina 201, Ciudad Empresarial, Huechuraba, Santiago, www.bisquertt.cl
Owner: Bisquertt family
Winemaker: Johana Pereira
Consulting winemaker: Felipe de Solminihac
Viticulturist: Jaime Araya
Known for: Cabernet Sauvignon, Carmenère, Syrah, Chardonnay, Gewürztraminer
Signature wine: La Joya
Other labels: Petirrojo, Ecos de Rulo, Tralca, Q Clay

Visiting: Open to the public; tasting room; restaurant and inn (Las Majadas)

In 1974, Osvaldo Bisquertt Reveco was one of the first to plant vines in Marchigue, twenty miles from the Pacific Ocean in coastal Colchagua. Johana Pereira, who has been with the winery since 1998, and Felipe de Solminihac are doing a tremendous job crafting wines from their 635 acres of vineyards in El Rulo and El Chequén. The lovely guesthouse has been beautifully remodeled and offers lunch and dinner.

Botalcura Winery

Founded: 2001
Address: Botalcura, Pencahue, Maule, www.botalcura.cl
Owner: Juan Fernando Waidele
Winemaker: Philippe Debrus
Viticulturist: Felipe Vial
Known for: Nebbiolo, Cabernet Sauvignon, Carmenère, Cabernet Franc
Signature wine: La Porfia
Other labels: Cayao
Visiting: Open to the public by appointment only; tasting room

Botalcura Winery is a project of two partners: the Chilean entrepreneur Juan Fernando Waidele and the French winemaker Philippe Debrus. Situated in western Chile's Maule, near the Coastal Mountains, it is adjacent to the town of Botalcura, northwest of Talca. Debrus, a Chilean resident since 1996 and former winemaker at Valdivieso, makes two ranges of wines and is one of the few to produce Nebbiolo

in Chile. *Botalcura* means "large stone" in Mapuche. The partners are extremely committed to sustainability, and the winery meets ISO standards for environmental management.

Calina

Founded: 1993
Address: Fundo El Maitén, Camino Las Rastras, km 7, Talca, Maule, www.calina.com
Owner: Jackson Family Wines
Winemaker: Randy Ullom, Andrés Sánchez
Viticulturist: Ricardo Marín
Known for: Chardonnay, Cabernet Sauvignon, Carmenère
Signature wines: Alcance, Bravura
Visiting: Open to the public by appointment only; tasting room

In 1999, the late Jess Jackson purchased the El Maiten estate, replanted the vineyards, and built a modern winery, designed by the architects Laurence Odfjell and John Purcell, about thirty miles east of Talca near the Lircay River. Calina has raised the profile of Maule while also pioneering other areas such as Limarí. Gillmore's Andrés Sánchez is involved in the day-to-day operation, and the Jackson Family Estates group enologist, Randy Ullom, has been involved since day one. Because grapes are sourced from diverse growers, the amount of estate wine in the Calina wines varies from vintage to vintage. A proposed new brand will be entirely from the home vineyard.

Caliterra

Founded: 1996

Address: Fundo Caliterra sin número,
El Huique, Santa Cruz, Colchagua,
www.caliterra.com

Owner: Eduardo Chadwick

Winemaker: Rodrigo Zamorano

Viticulturist: Ricardo Rodríguez

Known for: Carmenère, Cabernet Sauvignon, Syrah, Malbec

Signature wine: Cenit

Other labels: Tributo

Visiting: Tasting room by appointment only

The winery project encompasses 2,681 acres, 75 percent of which is untouched. In 2012, Viña Caliterra became the only winery to be listed in *Chile Verde* (Green Chile), a book promoting environmental culture that is a part of a larger project titled "Por un Chile Verde: Acciones para un Mundo Sostenible" (For a green Chile: Actions for a sustainable world), which supports best practices in environmental management. Viña Caliterra's sustainable-farming practices include efficient water use and recycling, reforestation, erosion-protection plans, the use of lightweight bottles, and encouraging grazing by wild horses to reduce the risk of fire. The winery was originally established with the Robert Mondavi winery in Napa Valley as a joint venture, but Mondavi was bought out in 2004. The name Caliterra combines the Spanish words for quality (*calidad*) and land (*terra*).

Canepa

Founded: 1930

Address: Peralillo, Colchagua Valley,
www.canepawinery.com

Owner: Familia Canepa, in a strategic alliance with Concha y Toro Group

Winemaker: Javier Solari

Known for: Cabernet Sauvignon, Carignan, Carmenère

Signature wine: Magnificum

Other labels: Genovino, Finísimo, Novísimo

Visiting: Tasting room by appointment only

Canepa was established in the port of Valparaíso by Giuseppe Canepa Vacarreza, from Genoa, Italy. Years later, this winery is a new entry into the Concha y Toro family that aims to make quality wines for everyday drinking. Its Genovino, made from sixty-year-old Carignan vines in Maule's Cauquenes region, is especially striking.

Carmen

Founded: 1850

Address: Av. Apoquindo 3669, Piso 7, Las
Condes, Santiago, www.carmen.com

Owner: Fundación Claro Vial (Santa Rita Group)

Winemaker: Sebastián Labbé

Consulting winemaker: Brian Croser

Viticulturist: Sebastián Warnier

Known for: Cabernet Sauvignon, Carmenère, Merlot, Syrah, Sauvignon Blanc, Chardonnay

Signature wine: Carmen Gold Reserve

Visiting: Open to the public by appointment only

Over 160 years old and named for the founder's wife, Carmen, this winery is among the oldest and most prestigious names in Chilean wine. It was a leader in Chile's early drive for quality wine and was the first to make a varietal Carmenère. It now sources fruit from an array of vineyards owned by the Claro Group, the parent company of the Santa Rita Group. Labbé's efforts to improve wine quality have made Carmen into a winery to watch.

Carolina Wine Brands

See Ochagavía, Santa Carolina, Viña Casablanca

South America's second-largest wine producer, Carolina Wine Brands is part of the leading Chilean agroindustrial group Watt's SA, whose primary shareholders are the Larraín family. The independently operated wine group, which has produced and sold high-quality wines for 135 years, owns seven winemaking facilities and some 4,940 acres of vineyards in the most prominent valleys of Chile and Argentina: Chile's Maipo, Casablanca, Colchagua, and Curicó, and Argentina's Mendoza.

The Antares brand is named for the brightest star in the constellation Scorpio, visible from the Southern Hemisphere. Its wines, another project of Santa Carolina, come in a wide range.

Carolina also owns Finca El Origen in the Uco Valley in Argentina, one of the first Argentinean wineries to arise from Chilean investment. The winery came into existence after several years of selling fruit to others and launched its first wines in 2002.

Casa Marín

Founded: 1999
Address: Camino Lo Abarca sin número, Lo Abarca, San Antonio, Aconcagua, www.casamarin.cl
Owner: María Luz Marín
Winemakers: Felipe Marín, María Luz Marín
Viticulturist: Felipe Marín
Known for: Sauvignon Blanc, Pinot Noir
Signature wines: Sauvignon Blanc Cipreses Vineyard, Pinot Noir Lo Abarca Hills
Other labels: Cartagena
Visiting: Open to the public; tasting room; guesthouse

María Luz Marín is one of Chile's great success stories, and her leadership in San Antonio is undisputed. Marín originally went to university to study agriculture but found herself bored out of her mind. She was about to quit, not sure what she wanted to do with her life, when she decided to take a chance on a viticulture class—and was hooked. She purchased eighty acres about three miles from the ocean. It contains a forest of eucalyptus trees that the Chilean government had designated as protected forest. After the government refused to budge, María had the trees yanked out in defiance. Just as the government was finally getting around to stopping her, the news media picked up the story, and she became a local celebrity, a *vinifera* revolutionary.

Casa Rivas

Founded: 1992

Address: Fundo Santa Teresa sin número, María Pinto, Maipo, www.casarivas.cl

Owner: VSPT Wine Group

Winemaker: Deborah Witting

Viticulturist: Tomás Rivera

Known for: Cabernet Sauvignon, Carmenère, Syrah

Signature wine: Gran Reserva Carmenère

Visiting: Open to the public; tasting room

Founded by two Chilean wine viticulturists, Mariano Salas Rivas and Patricio Browne, this Maipo-based winery joined the VSPT group in 2005. Based in María Pinto, it makes a full range of wines.

Casas del Bosque

Founded: 1993

Address: Hijuelas No. 2, ex Fundo Santa Rosa, Casablanca, Aconcagua, www.casasdelbosque.cl

Owner: Juan Cúneo Solari

Winemaker: Grant Phelps

Consulting winemakers: Ken Bernards, Pilar González

Viticulturist: Carlos Peña

Known for: Sauvignon Blanc

Signature wine: Gran Estate Selection

Other labels: Pequeñas Producciones

Visiting: Open to the public; tasting room; restaurant (Tanino)

Immense woods of ancient pine and olive trees and small, abandoned white adobe houses were key in determining the location of this vineyard and manor house, giving the property and winery the name Casas del Bosque, or "houses of the forest." There are 618 acres of vineyards, 296 on hills and the balance on the valley floor. The winery also owns vineyards in Maipo that provide fruit for its Carmenère and Cabernet Sauvignon. The wildly energetic New Zealand winemaker Grant Phelps arrived in 2010 and has dramatically improved quality. Though renowned for Sauvignon Blanc, Casas del Bosque also produces remarkable cool-climate Syrahs and operates a great award-winning destination restaurant, Tanino.

Casas del Toqui

Founded: 1994

Address: Fundo Santa Anita de Totihue sin número, Requínoa, Cachapoal, www.casasdeltoqui.cl

Owner: Granella family and Allianz Group (Château Larose Trintaudon)

Winemaker: Alfonso Duarte Piña

Consulting winemaker: Sergio Correa Undurraga

Known for: Cabernet Sauvignon, Merlot, Syrah, Chardonnay, Carmenère, Pinot Noir, Sauvignon Blanc

Signature wine: Gran Toqui

Other labels: Leyenda del Toqui

Visiting: Open to the public by appointment only

What started as a joint venture between Château Larose Trintaudon of Bordeaux and a respected Chilean vineyard family in Totihue has quickly established itself as a high-quality producer. Its vineyards are mostly located in Alto Cachapoal, about

250 miles south of Santiago, and the late-harvest Semillon is considered one of Chile's best dessert wines.

Casa Silva

Founded: 1997

Address: Hijuelas Norte sin número, Angostura, San Fernando, Colchagua, casasilva.cl

Owner: Mario Silva

Winemakers: Mario Geisse, José Ignacio Maturana

Viticulturist: Mario Geisse

Known for: Carmenère, Cabernet Sauvignon, Sauvignon Blanc

Signature wine: Microterroir de Los Lingues Carmenère

Visiting: Open to the public; tasting room; restaurant and hotel

This winery has made a dramatic turnaround since the 1990s, when it was known more for volume than for quality. Specializing in Carmenère, with signature wines that are among Chile's best examples, Casa Silva has been involved with a research project on microterroir directed by Yerko Moreno at the University of Talca. The goal of identifying soils and pairing them up with the best varieties and selections has been very successful. The published studies of Casa Silva's celebrated Los Lingues vineyard have aided other wineries embarking on such efforts. Mario Geisse, who owns a namesake sparkling-wine domain in Brazil, oversees production. The seven-room guesthouse is magnificent.

Catrala

Founded: vineyards (1994); winery (2003)

Address: Fundo Don Manuel, Camino Lo Orozco sin número, km 10, Casablanca, Aconcagua, www.catrala.cl

Owner: Rodríguez family

Winemaker: Ana María Pacheco

Viticulturist: Felipe Rodríguez

Known for: Pinot Noir, Sauvignon Blanc

Signature wine: Pinot Noir Grand Reserve Limited Edition

Visiting: Open to the public; tasting room

The winery was named for Catrala, a locally famous and enigmatic seventeenth-century Chilean woman known for her eccentricity, elegance, and impulsiveness. The facility, based in Lo Orozco, works from 173 acres of vineyard.

Clos des Fous

Founded: 2009

Address: Apoquindo 3500, Piso 12, Las Condes, Santiago, www.closdesfous.com

Owners: Albert Cussen, Francisco Leyton, François Massoc, Pedro Parra

Winemaker: François Massoc

Viticulturists: Francisco Leyton, Pedro Parra, François Massoc

Known for: Pinot Noir, Chardonnay, Riesling, Cabernet Sauvignon

Signature wine: Clos des Fous Pinot Noir

Visiting: Open to the public by appointment only

Clos des Fous (the "madmen's vineyard" in French) is an exciting effort by François Massoc and Pedro Parra, two longtime friends who are brilliant experts in their respective disciplines: Parra in viticulture and Massoc in winemaking. Massoc, who has made wine in Bolivia, Palestine, and Israel and works on myriad other projects (including Aristos), is one of Chile's best winemakers, and Parra is nearly deified. Together they seek sites that are far from mainstream and are producing amazing bottles from areas as improbable as Traiguén in the southern extreme of Malleco to very old vines in Itata. Their approach produces wines that distinctively express their terroir, made with organic fruit and using techniques that are so minimalist that the wines are vinified without "safety nets" such as fining or filtering.

Concha y Toro

Founded: 1883

Address: Av. Nueva Tajamar 481, Torre Norte, Piso 15, Las Condes, Santiago, www.conchaytoro.com, www.trioconchaytoro.com

Owners: Publicly held. Principal owners: Guilisasti family (26.8%) and Alfonso Larraín (7.1%).

Winemakers: Ignacio Recabarren, Enrique Tirado, Marcelo Papa

Viticulturist: Andrés Larraín (agricultural manager)

Consulting winemaker and viticulturist: Eric Boissenot

Known for: Cabernet Sauvignon, Carmenère, Merlot, Chardonnay, Sauvignon Blanc, Syrah

Signature wines: Don Melchor, Casillero del Diablo, Carmín de Peumo

Visiting: Open to the public at the Pirque Wine Center (Av. Virginia Subercaseaux 210, Pirque, Maipo Valley); tasting rooms at the Pirque Wine Center and in the Tasting Center in Santiago (Av. Alonso de Córdova 2391, Vitacura, Santiago); restaurant

The Concha y Toro winery is the largest of six Chilean wineries operated by the parent company. It accounts for about 25 percent of all Chilean wines sold. It is, of course, the largest Chilean wine producer, selling some three million cases of its primary line of Casillero del Diablo wines worldwide. It manages to combine high volume (13.6 million cases annually) and admirable quality, offering good wine across several price ranges as well as superb super-premium offerings: its celebrated "Don Melchor" is among Chile's very best. The dynamic duo of Ignacio Recabarren and Marcelo Papa oversees this colossal effort with aplomb.

Concha y Toro Group

See Almaviva, Canepa, Concha y Toro, Cono Sur, Maycas del Limarí, Viña Maipo

It is hard to summarize the importance of the Concha y Toro Group in a single paragraph. As of December 2009, the company had approximately 27,059 acres on forty-one company-owned and eight leased vineyards in the Limarí, Aconcagua, Casablanca, San Antonio, Maipo, Cachapoal, Colchagua, Curicó, and Maule growing areas in Chile and eight vineyards in Mendoza, Argentina. It is the single largest wine group in South America, with a strong commitment to quality.

Cono Sur

Founded: 1993

Address: Av. Nueva Tajamar 481, Torre Sur, Piso 21, Oficina 2101, Las Condes, Santiago, conosur.com

Owner: Concha y Toro Group

Winemaker: Adolfo Hurtado

Viticulturist: Gustavo Amenábar

Known for: Pinot Noir

Signature wine: Ocio

Visiting: Open to the public by appointment only

Cono Sur, named for its location in the Southern Cone of South America, was the first winery to be recognized as being carbon neutral, in 2007. This winery first caught my attention in successive years of blind judging at the Korea Wine Challenge in Seoul, where it won "best of class" awards in four consecutive years. It is one of the most consistent and reliable wineries in the country. Half of the Colchagua estate grapes are farmed organically, as are 25 percent of all the grapes from the eight estates and fifteen growers with whom the winery works. Acknowledged as the national leader in Pinot Noir, which represents nearly as much of its production as Cabernet Sauvignon, Cono Sur, led by the accomplished Adolfo Hurtado, also bottles distinctive Riesling and Viognier and high-quality sparkling wine.

Cousiño Macul

Founded: 1856

Address: Av. Quilín 7100, Peñalolén, Santiago, www.cousinomacul.com

Owner: Cousiño family

Winemaker: Pascal Marty

Viticulturist: Carole Dumont

Known for: Cabernet Sauvignon, Merlot, Chardonnay

Signature wines: Lota, Antiguas Reservas Cabernet Sauvignon

Other labels: Finis Terrae

Visiting: Open to the public by appointment only; tasting room

The first South American wines I ever tasted were from this historically significant operation. It's the only nineteenth-century Chilean winery that remains in the hands of the founding family. Having struggled to keep pace with the national revolution in wine, the sixth generation of owners, in 1996, decided to cash in on rising land prices by developing and selling off much of its property in the area and keeping only the family home and adjoining 154-acre park. After deciding to continue producing wine, they acquired 815 acres twenty miles away, in Buin in southwest Maipo, of which they planted 679 acres in grapes. The new Cousiño Macul is still a work in progress, but the direction is positive, and Pascal Marty's experiences with the Baron de Rothschild's French estates should aid the effort.

Cremaschi Furlotti

Founded: 1889

Address: Estado 359, Piso 4, Santiago, www.cf.cl

Owner: Pablo Cremaschi

Winemaker: Rodrigo González

Viticulturist: Cristian Cremaschi

Known for: Cabernet Sauvignon, Carmenère, Syrah, Nebbiolo, Sauvignon Blanc, Chardonnay, Pinot Grigio

Signature wine: Vénere

Visiting: Open to the public by appointment only; tasting room

The Cremaschi Furlotti family emigrated from Italy four generations back. The third-generation paternal grandfather migrated to Argentina and founded the formidable Viña Furlotti. The family that remained in Chile continued the winemaking tradition in Maule's Loncomilla and now makes wines from the Peñasco vineyard under the Tierra del Fuego label. Rodrigo González has been the winemaker since 2000.

Dalbosco

Founded: 2008

Address: Cienfuegos 361, La Serena, Coquimbo, www.dalboscowines.com

Owners: Franco, Giorgio, Renato, and Roberto Dalbosco Cazzanelli

Winemaker: Lorena Veliz Angel

Consulting winemaker: Felipe de Solminihac

Viticulturists: Guillermo Madrid, Felipe de Solminihac

Known for: Syrah, Carmenère, Merlot, Viognier, Cabernet Sauvignon

Signature wine: Dalbosco Assemblage Reserve

Visiting: Open to the public by appointment only

The Dalbosco Cazzanelli family emigrated from the Trentino region of Italy in 1952. Almost sixty years later, Viña

Dalbosco was founded by the second-generation brothers in the Limarí Valley's rural Punitaqui, making wine from vineyards planted by their father.

De Martino

Founded: 1934

Address: Manuel Rodríguez 229, Isla de Maipo, Maipo, www.demartino.cl

Owner: De Martino family

Winemakers: Marcelo Retamal, Eduardo Jordan

Viticulturists: Héctor Bertero, Renán Cancino

Known for: Carmenère, Cabernet Sauvignon, Merlot, Chardonnay, Sauvignon Blanc, Syrah, Carignan, Malbec, Cinsault

Signature wine: De Martino Familia

Other labels: Organic, Legado, 347 Vineyards, Viejas Tinjas, Vigno

Visiting: Open to the public; tasting room; restaurant

The home base of the esteemed Marcelo Retamal, one of Chile's most accomplished winemakers and known to his colleagues as *el doctor,* De Martino is considered to be one of Chile's top terroir- and site-focused wineries. Retamal, an authentic craftsman, takes a minimalist approach. He uses no new oak, preferring large older casks, and promotes the use of the old ceramic *tinjaras,* clay amphorae, for fermentations. There are no cultivated yeasts, no filtration, and no intervention. De Martino has sleuthed sites in previously unrecognized wine areas like Choapa and Elquí in search of cooler-climate spots.

Domaine de Manson

Founded: 2000

Address: Hijuela 11, Coirón, Salamanca, Coquimbo; no website

Owner: Arnaud Faupin

Winemaker: Arnaud Faupin

Viticulturist: Arnaud Faupin

Known for: Syrah, Carmenère, Viognier

Signature wine: Turca

Visiting: Open to the public by appointment only

At less than ten acres, this is probably Chile's smallest estate vineyard. Having bought the whole enterprise—vines, brand, and barrels—from George Manson, Arnaud Faupin plans to stay the course. The winery is located about five kilometers east of the village of Coirón, in the commune of Salamanca, between Coirón and the banks of the Choapa river. The small Rhône-style winery is a member of the MOVI association.

Dos Andes

See Gracia de Chile, Llai Llai, Porta

Formerly known as Viña Córpora, Dos Andes (www.dosandeswines.com) is the leading producer in the Bío-Bío region of Chile, with close to one thousand acres spread around Negrete, Mulchén, Yumbel, and Quinel. The company owns 3,700 acres of vineyards throughout Chile and additional plantings in Argentina, stemming from the 2008 purchase of Bodega Universo Austral winery and including more than 700 acres of vineyards that straddle the provinces of Neuquén and Río

Negro. It also owns the Argentinean wineries Finca Roja, Ruta 22, and Calafate.

Echeverría

Founded: 1930

Address: Viñedos la Estancia sin número, Molina, Curicó, www.echewine.com

Owners: Roberto Echeverría P. and family

Winemakers: Roberto Echeverría Z. (chief winemaker), Matías Aguirre (resident winemaker)

Viticulturist: Cecilia Silva

Known for: Cabernet Sauvignon, Carmenère, Pinot Noir, Syrah, Sauvignon Blanc, Chardonnay

Signature wine: Limited Edition Cabernet Sauvignon

Visiting: Open to the public by appointment only; tasting room

In 1740, the Echeverría family moved from Amezqueta in the Basque country to Chile, where they became farmers, grape growers, and winemakers. In 1923, they purchased a 200-acre property in the town of Molina in the Curicó Valley, which they planted with vines brought from France, and began making bulk wine from *vinifera* grapes at a time when the region was awash with local criolla wine production. In 1978, when the grandfather of Roberto Echeverría passed away, the family was living in the United States but decided to return to Chile. Roberto's father worked as a banker during the week and as a winemaker on weekends, putting his earnings back into the winery. Echeverría is a traditional winery and augments its estate grapes with fruit purchased from areas such as Maipo and Colchagua.

El Principal

Founded: 1992

Address: Fundo El Principal sin número, Pirque, Maipo, www.elprincipal.cl

Owner: Döhle Latinamerika SA

Winemaker: Gonzalo Guzmán

Consulting winemaker: Patrick Leon

Known for: Cabernet Sauvignon, Carmenère

Signature wine: El Principal

Other labels: Memorias, Calicanto

Visiting: Open to the public by appointment only; tasting room

El Principal winery was originally a partnership between Jorge Fontaine, owner of the Hacienda El Principal, and the late Jean-Paul Valette, former owner of Château Pavie in Saint-Émilion and father of the eminent Chilean-French winemaker Patrick Valette of Vik. Their objective was to produce top-quality red wines in the French tradition from a 133-acre property in Maipo. A few years after Jean-Paul's passing, the current owners bought the winery operation and vineyard. El Principal is the easternmost winery in Chile, fifteen minutes east of Pirque, in the heart of the Andean piedmont.

Emiliana

Founded: 1986

Address: Av. Nueva Tajamar 481, Torre Sur, Oficina 905, Las Condes, Santiago, www.emiliana.cl

Owners: Shareholders (majority Guilisasti family)

Winemaker: César Morales

Consulting winemaker: Álvaro Espinoza

Viticulturist: Miguel Elissalt

Known for: Cabernet Sauvignon

Signature wine: Coyam

Visiting: Open to the public; tasting room

The Guilisasti brothers, José and Rafael, are substantial shareholders in Concha y Toro. In the late 1990s they were among the first to realize that wine consumers worldwide were becoming more concerned about the social and environmental conditions under which wine was produced. After more than a decade, they have realized their dream. Half of their grapes are certified organic and biodynamic, making it the largest operation of its kind in South America. Their "Gê" is Chile's first-ever certified biodynamic wine.

Errázuriz

Founded: 1870

Address: Calle Antofagasta sin número, Panquehue, Aconcagua, www.errazuriz .com

Owner: Eduardo Chadwick

Winemaker: Francisco Baettig

Consulting winemaker: Louis-Michel Liger-Belair

Viticulturist: Jorge Figueroa

Known for: Cabernet Sauvignon, Carmenère, Syrah

Signature wine: Don Maximiano Founder's Reserve

Other labels: La Cumbre, Kai

Visiting: Open to the public by appointment only; tasting room

The work of the visionary Eduardo Chadwick over the past ten years or more has been revolutionary in Aconcagua and beyond. Since 1993 he's been president of Viña Errázuriz, which has been a family business since its founding, except for a brief interlude in the 1970s and 1980s when it was owned by a bank. He is constantly pushing for improvement throughout the winery's range, and many of his wines are among Chile's very best. His iconic wines showed up well against French and Tuscan "first growths" in the Berlin Tasting in 2004 and have also performed well in fourteen other competitions over seven years, giving international cachet to his super-premium wines.

Estampa

Founded: 2001

Address: Carretera del Vino, km 47, Palmilla, Colchagua, www.estampa.com

Owner: Miguel González

Winemaker: Johana Pereira

Consulting winemaker: Attilio Pagli

Viticulturist: Martín Fernández

Known for: Carmenère, Syrah, Cabernet Sauvignon, Viognier, Sauvignon Blanc

Signature wine: LaCruz

Other labels: Gold, Delviento

Visiting: Open to the public; tasting room

The name "Estampa" originally referred to the family mill. When Miguel González's grandfather set up the winery, he retained the name to preserve the link with the family's past. Although the winery is in Palmilla, Estampa sources grapes throughout Colchagua to produce its wines and has been a pioneer in the cool-climate area of Paredones, near the coast. Until spring 2014, Estampa's winemaking was headed by Marcela Chandia, one of the gifted young female winemakers so prominent in the contemporary Chilean wine scene. She was replaced by Johana Pereira, who also oversees the team at Bisquertt.

Falernia

Founded: 1998

Address: Ruta 41, km 52, Casilla 8, Vicuña, Coquimbo, www.falernia.com

Owners: Olivier family, Giorgio Flessati

Winemaker: Giorgio Flessati

Viticulturist: Eduardo Silva

Known for: Syrah, Carmenère

Signature wine: Carmenère Reserve

Other labels: Donna Maria, Number One

Visiting: Open to the public; tasting room

In 1972, Aldo Olivier moved his family to Elquí, where he planted grapes for pisco production. He became vice president of the pisco cooperative, but he soon recognized that the co-op world was not for him, and in 1975 he started his own pisco project and constructed a winery. In 1995, Aldo's cousin from Trentino, the Italian winemaker Giorgio Flessati, visited the valley, and the two decided to make wine together. Their focus is on Syrah, Sauvignon Blanc, Carmenère, and the two pisco grapes Pedro Ximénez and Torontel (Torrontés). Many aficionados consider their Syrah to be the best in the country!

Mayu: Mayu (www.mayu.cl) was started as an independent winery within the Olivier family group in 2005, run by Mauro Olivier Alcayaga. The wines are very similar in style, with winemaking also overseen by Giorgio Flessati.

Flaherty

Founded: 2004

Address: Casilla 415, San Felipe, Aconcagua, www.flahertywines.com/chile/welcome/

Owners: Ed Flaherty, Jen Hoover

Winemaker: Ed Flaherty

Viticulturist: Ed Flaherty

Known for: Syrah, Cabernet Sauvignon, Tempranillo

Signature wine: Flaherty

Visiting: Open to the public by appointment only

Ed Flaherty and Jen Hoover came to Chile from the United States in 1993 to work a harvest and never left. Avid kayakers, they claim that Chile's great whitewater had more than a little to do with their decision to stay. These days, they focus on wine. Ed has experience with several of Chile's top wineries, including Cono Sur, Errázuriz, Vía Wines, and now Viña Tarapacá. In 2008, they bought thirty-five acres in Cauquenes (Maule). Jen is an avid chef and writes an informative and enticing food blog for their website.

Garage Wine Company

Founded: 2009

Address: Camino San Antonio, Caliboro, km 5.8, San Javier, Maule, www.garagewineco.cl

Owners: Derek Mossman Knapp, Pilar Miranda

Winemakers: Pilar Miranda, Derek Mossman Knapp

Known for: Cabernet Sauvignon, Carignan

Signature wine: Old Vine Carignan

Visiting: Open to the public by appointment only

The enthusiastic couple makes about 1,500 cases annually of Carignan, Cabernet Franc, Cabernet Sauvignon, and Pinot Noir sourced from small growers in Maule and Maipo. Advocates of minimal intervention, they make their wines *muy natural:* fermented only with native yeast, with no enzymes or fining additions of any kind, and with very little new wood. If you buy more than one bottle, you may notice slight variations in bottle thickness, since Mossman Knapp takes recycling full circle and reuses salvaged wine bottles, in partnership with a recycling company near the winery. Though he claims no credit, he is the founder of MOVI, Chile's exciting, fast-growing association of independent vintners, and Vigno, the admirable old-vine Carignan project based in Maule.

GEO Wines (Chono)

Founded: 2001

Address: Av. Nueva Providencia 1860, Oficina 92, Providencia, Santiago, www.geowines.cl

Owner: Empresas Sutil (Juan Sutil)

Winemaker: Juan Carlos Faúndez

Consulting winemaker: Álvaro Espinoza

Known for: Sauvignon Blanc, Chardonnay, Pinot Noir, Cabernet Sauvignon, Carmenère, Syrah

Signature wine: Chono Reserva

Other labels: Cucao, Rayun, Vilcun

Visiting: Open to the public by appointment only

GEO wines is a collaboration of two of Chile's great winemaking talents, Álvaro Espinoza and Juan Carlos Faúndez. In addition to making their own wines, an effort they began just over a decade ago, both are respected wine consultants. Juan Carlos consults for Viña Pérez Cruz, Casa Rivas, Emiliana, Ventolera, TerraMater, and Apaltagua. GEO sources fruit from four different regions and is committed to sustainability.

Gillmore

Founded: 1990

Address: Camino a Constitución, km 20, San Javier, Maule, www.gillmore.cl

Owner: Francisco Gillmore

Winemaker: Andrés Sánchez

Viticulturist: Daniella Gillmore

Known for: Cabernet Franc, Cabernet Sauvignon, Merlot, Carignan

Signature wine: Hacedor de Mundos

Other labels: Cobre

Visiting: Open to the public; tasting room; restaurant (La Cave de Francisco); lodge (Tabonkö Guesthouse)

After Francisco Gillmore bought the property, he grafted premium grapes onto old Pais vines, and we are all the beneficiaries of that change. In 2002 his daughter Daniella took over the winery in conjunction with her husband, the talented winemaker Andrés Sánchez. Beyond the vines, which cover about 10 percent of the thousand-acre property, the Gillmores' restaurant and wine-themed guest lodge are well worth a visit. A portion of the property is dedicated to preserving Chile's indigenous wildlife: in its nature preserve and mini-zoo are parrots, geese, and mountain lions. Gillmore offers a five-day winemaking course for groups of six or more.

Gracia de Chile

Founded: 1989

Address: Av. Las Condes 11.380, Piso 8, Vitacura, Santiago, www.gracia.cl

Owner: Dos Andes

Winemakers: Louis Vallet, Marcela Chandia

Consulting winemaker: Pascal Marchand

Viticulturist: Carlos Carrasco

Known for: Carmenère, Cabernet Sauvignon, Sauvignon Blanc, Chardonnay, Merlot, Pinot Noir, Syrah, Malbec

Signature wine: Gracia de Chile Sauvignon Blanc

Other labels: Veranda, Agustinos

Visiting: Open to the public by appointment only

Gracia de Chile is a joint venture between Dos Andes and Boisset Family Estates, the large wine conglomerate in France and California. Winemaking is overseen by the very capable team of Louis Vallet, the son of Bernard Vallet of the eponymous Burgundy domain, and Marcela Chandia, the winemaker of Viña Estampa in Colchagua. The white wines are blends from fruit grown across Dos Andes's Bío-Bío properties; the Pinot Noirs are estate specific. The other red wines come from vineyards in Colchagua and Aconcagua.

Hacienda Araucano

Founded: 1997

Address: Ruta I-72, km 29, Lolol, Santa Cruz, Colchagua, www.francoislurton .com

Owner: François Lurton

Winemakers: François Lurton, Luca Hodgkinson

Viticulturists: François Lurton, Luca Hodgkinson

Known for: Cabernet Sauvignon, Sauvignon Blanc, Pinot Noir, Carmenère, Syrah, Cabernet Franc

Signature wine: Alka

Other labels: Gran Araucano, Clos de Lolol, Humo Blanco, Kawin

Visiting: Open to the public by appointment only; tasting room

The Chilean Lurton-owned effort is based primarily on grapes emanating from its own vineyards, consisting of nearly sixty-five acres in the cooler Lolol zone, twenty-five miles from the ocean. A thirty-eight-acre vineyard farther down in the Colchagua Valley provides fruit for the Carmenère and Cabernet Sauvignon. The wines are definitively Old World and lean in personality. The vineyards, which have been cultivated according to biodynamic principles since 2009, have been certified organic for the 2013 vintage.

Haras de Pirque

Founded: 2000

Address: Fundo La Rochuela sin número, Pirque, Maipo, www.harasdepirque.com

Owners: Eduardo Matte, Piero Antinori

Winemaker: Cecilia Guzmán

Consulting winemaker: Renzo Cotarella

Viticulturist: Andrés Aparicio

Known for: Cabernet Sauvignon, Carmenère, Cabernet Franc, Syrah, Chardonnay, Sauvignon Blanc

Signature wine: Haras Elegance

Other labels: Equus, Haras Character, Albis

Visiting: Open to the public by appointment only; tasting room

Haras de Pirque claims that its eye-catching, horseshoe-shaped winery is "the only one of its kind in the world." I suspect they are right about that! Haras de Pirque is named for the country's oldest thoroughbred racehorse stud farm, founded in 1892. The farm has four extensively pedigreed stallions that have sired some exceptional offspring. The winery came into existence when Eduardo Matte bought the estate, including the stud farm, on the western edge of Pirque. Close to three hundred acres of vines were planted in the early 1990s, and the remaining land has been used for the horses (the farm is equipped with a training center and a 1,500-meter track) and for fruit plantations. A collaboration with Piero Antinori in 2002 resulted in "Albis," a *corte* of Cabernet Sauvignon and Carmenère.

Henriquez Hermanos Ltda (Viña El Aromo)

Founded: 1922

Address: Diecisiete Oriente 931, Talca, Maule, www.elaromo.com

Owner: Henriquez family

Winemakers: Jimena Egaña, Fernando Torres, Ignacio Saavedra

Viticulturist: Carlos Mondaca

Known for: Cabernet Sauvignon, Carmenère, Merlot, Malbec, Petit Verdot, Syrah, Sauvignon Blanc, Chardonnay

Signature wine: Aromo Barrel Selection

Other labels: Dogma, Casa Lo Matta

Visiting: Open to the public by appointment only

El Aromo was acquired in 1940 by Victor Henriquez, a grape grower in Talca. Today it is under the management of his sons, Arturo and Manuel. They work from 707 acres spread across four vineyards in the Maule Valley and produce wines under three separate labels.

J. Bouchon

Founded: 1892

Address: Mingre, Camino a Constitución, km 30, San Javier, Maule, www.jbouchon.cl

Owner: Julio Bouchon Sepúlveda

Winemaker: Rafael Sánchez

Consultant: Patrick Valette

Viticulturist: Pablo Toro

Known for: Malbec, Cabernet Sauvignon, Syrah, Sauvignon Blanc

Signature wine: Mingre

Other labels: Las Mercedes, Convento Viejo

Visiting: Open to the public by appointment only; tasting room; restaurant

Named for the owner, Julio Bouchon, the winery has a history going back to the end of the nineteenth century, when Émile Bouchon, like so many, left a French estate ravaged by phylloxera and arrived in Chile's Maule Valley. Third-generation Julio Bouchon gave the winery its renown. Like many other wineries in Maule, Bouchon was hard hit by the 2010 earthquake and has since been rebuilt. Located near the coast, it has about 915 acres of old vines across a number of properties in Santa María de Mingre, Santa Rosa, Las Mercedes, and Batuco.

Kankura

Founded: 2005

Address: Fundo La Ensenada sin número, Palmilla, Colchagua, www.kankura.cl

Owners: Fernando Ovalle, José Hernán Ovalle

Winemaker: Raúl Narváez

Viticulturist: Raúl Narváez

Known for: Malbec, Syrah, Cabernet Sauvignon

Signature wine: Malbec Édition Limitée

Visiting: Open to the public by appointment only

Kankura, an indigenous word for "jar of clay," is symbolic of the winery's commitment to traditions of the land and the craftsmanship of creating a great work from simple raw materials. Formerly known as Ensenada de Palmilla, the "amphitheatre of Palmilla" (for its location, encircled by mountains), the 183-acre estate was purchased in 1976 by two lifelong (unrelated) business partners and friends, José Hernán Ovalle and Fernando Ovalle. Kankura was born when they hired the renowned enologist Raúl Narváez to head up their winery.

Kingston Family Vineyards

Founded: 2003 (first vintage)

Address: Hijuela El Maitén, Casablanca, Aconcagua, www.kingstonvineyards.com

Owner: Kingston family

Winemaker: Evelyn Vidal

Consulting winemaker: Byron Kosuge

Viticulturist: Patricio Monsalva

Known for: Pinot Noir, Syrah, Sauvignon Blanc

Signature wine: Bayo Oscuro Syrah

Other labels: Alazan, Tobiano, Lucero, Cariblanco

Visiting: Open to the public by appointment only; tasting room

Michigan-born Carl John Kingston arrived in Chile in the early 1990s in search of precious metals and acquired a 12,000-acre estate in coastal Casablanca. The mining enterprise turned out to be a bust, but the property has remained with the family since, although most of them now live in the United States. The veteran enologist Byron Kosuge was formerly winemaker at Saintsbury. Kingston originally sold most of its fruit, partly as a way of advertising its quality to grape buyers, and didn't make its own wines until quite recently. These have become known as among Chile's best. The winery intends to expand slowly and to use a maximum of 10 to 15 percent of the vineyard fruit for the estate wines.

Koyle

Founded: 2006

Address: Isidora Goyenechea 3600, Oficina 1101, Las Condes, Santiago, www.koyle.cl

Owner: Undurraga family

Winemaker: Cristóbal Undurraga

Viticulturist: Cristóbal Undurraga

Known for: Syrah, Cabernet Sauvignon, Carmenère, Malbec, Sauvignon Blanc

Signature wine: Royale

Visiting: Open to the public by appointment only

Named after an endangered native Chilean plant, this premium winery from the Colchagua Valley is a new project from a familiar name. After leaving the family winery following a dispute, Alfonso Undurraga Mackenna, a great-nephew of the founder, decided to pursue his dream of finding a superior terroir from which he and his three sons could create small lots of hand-crafted wines. They settled on a large parcel in the Los Lingues zone of Alto Colchagua. His son Cristóbal spent a year at Château Margaux under winemaker Paul Pontallier, who had earlier worked under Alfonso (it's a small world). And then Cristóbal worked at Viña Montes under Aurelio Montes a few years later, again because Aurelio had worked under his father. This winery is generating lots of excitement and operates according to biodynamic principles.

Laberinto

Founded: 1996

Address: Ribera Sur Lago Colbún, km 14, Maule, www.laberintowines.cl

Owners: Georg Andresen, Rafael Tirado

Winemaker: Rafael Tirado

Viticulturist: Rafael Tirado

Known for: Sauvignon Blanc

Signature wine: Cenizas de Barlovento Sauvignon Blanc

Visiting: Open to the public; tasting room

The thirty-five-acre vineyard is a father/son-in-law project that started in 1993 with a single acre of Cabernet Sauvignon and has slowly evolved. Rafael Tirado started in wine at neighboring TerraNoble and subsequently was appointed enologist at Casablanca's Veramonte, where, through the Huneeus family, he also participated in several Napa Valley vintages. The name of the winery, which means "labyrinth," alludes to the unique maze-like planting of the vines, in contrast to the traditional measured and spaced rows. Believing that cultivating vines in perfectly uniform rows falsely assumes that all the grapes receive the same wind, soil, water, and sun exposure, Laberinto's team is convinced that the circular shape of the vineyards adds complexity of aromas and flavors to the wine. Their signature Sauvignon Blanc is a brilliant wine.

La Playa

Founded: 1987

Address: Camino a Calleuque sin número, Peralillo, Colchagua, www.laplayawine.com

Owners: Axelsen family, Sutil family, Errázuriz family

Winemaker: Oscar Salas

Consulting winemaker: Rafael Tirado

Viticulturist: Rodrigo Serrano

Known for: Cabernet Sauvignon, Merlot, Carmenère, Pinot Noir, Claret, Sauvignon Blanc, Chardonnay, Syrah

Signature wine: Axel Primero

Other labels: Axel, Loud River Roaring Red

Visiting: Open to the public by appointment only; tasting room; restaurant and hotel (Hotel Viña La Playa)

The Axelsen family came to Chile from California's Marin County. Rather than follow the traditional business model of purchasing land, building a winery, and launching a brand, they started out by custom crushing their wines in other producers' facilities, with the eventual goals of purchasing vineyards and building their own winery. They partnered with the Chilean families to acquire 597 acres of prime property in Colchagua and constructed a modern winery with the latest in technology. The bodega also has an elegant four-star hotel nestled in the vineyards—the first winery-based hotel in Chile—with eleven rooms, tennis courts, and a heliport. The Axelsen family owns Cabernet Corporation, a U.S. importer based in Marin that specializes in Chilean and Argentinean brands.

Lapostolle

Founded: 1994

Address: Ruta I-50, Camino San Fernando a Pichilemu, km 36, Cunaquito, Santa Cruz, Colchagua, www.lapostolle.com

Owners: Alexandra Marnier Lapostolle, Cyril de Bournet

Winemakers: Jacques Begarie, Andrea León

Consulting winemaker: Michel Rolland

Viticulturist: Jorge Castillo

Known for: Sauvignon Blanc, Chardonnay, Merlot, Carmenère, Cabernet Sauvignon, Syrah

Signature wine: Clos Apalta

Other labels: Cuvée Alexandre, Casa, Canto de Apalta, Borobó

Visiting: Open to the public; restaurant; hotel (Lapostolle Residence)

The Lapostolle winery is one of the most significant in Chile. It was originally established as a joint venture between the Chilean Rabat family and Alexandra Marnier Lapostolle, whose family owns the Grand Marnier liqueur operation as well as Château de Sancerre in the Loire. Lapostolle acquired the Rabats' holding less than ten years later. From the beginning, the winery hired the French consultant winemaker Michel Rolland, on terms that secured exclusive rights to his services in Chile. Although it is hard to overstate the role of this winery in launching the modern era of Chilean wine, it has only continued to improve. Lapostolle's luxury accommodations and dining are sublime, though not designed for those on a budget.

Laura Hartwig

Founded: 1994

Address: Camino Barreales sin número, Santa Cruz, Colchagua, www.laurahartwig.cl

Owner: Comercial Santa Laura SA

Winemaker: Renato Czischke

Consulting winemaker: Felipe García

Viticulturist: Alejandro Hartwig

Known for: Cabernet Sauvignon, Carmenère, Merlot, Petit Verdot, Malbec, Cabernet Franc, Syrah

Signature wine: Gran Reserva Blend

Visiting: Open to the public; tasting room; restaurant (Vino Bello); hotel (Terraviña)

Laura Hartwig's father purchased the land in 1928 and gave it to her in 1966. When the family moved to Canada, it lay vacant for years. In the 1970s Laura told her winemaker husband that she would agree to move back to Chile only if he agreed to make Chardonnay, her favorite wine. The Hartwigs have kept the winery a low-key, boutique-style affair, producing wines mainly for restaurants and for sale at the winery. The Terraviña hotel, with nineteen well-appointed rooms, a polo field, stables, and restaurant, is a convenient attraction.

Leyda

Founded: 1997

Address: Camino a Cuncumen sin número, km 4, Leyda, Aconcagua, www.leyda.cl

Owner: VSPT Wine Group

Winemaker: Viviana Navarrete

Viticulturist: Cristóbal Mujica

Known for: Pinot Noir, Sauvignon Blanc, Syrah

Signature wine: Lot 21 Pinot Noir

Other labels: Lot, Single Vineyard

Visiting: Open to the public by appointment only; tasting room

In 2007, Leyda was acquired by Viña Tabalí, which is owned jointly by the San Pedro group and Guillermo Luksic. Most of my friends and associates in the wine business consider this winery to be one of the most exciting in all of Chile. It is the pioneering winery in the Leyda region: for a time, the original vineyard, planted in 1998, *was* the Leyda region. Now things are a bit more crowded, with over 2,200 acres of nearby vineyards. The original vineyard is on a 280-acre property, with 215 acres planted to vine. A more recent 223-acre acquisition, El Granito, is even closer to the ocean. Viña Leyda's various Sauvignon Blancs are stunning, as are their ranges of Chardonnay and Pinot Noir.

Llai Llai

Founded: 2010

Address: Av. Las Condes 11.380, Piso 8, Vitacura, Santiago, www.llaillai.com (website under construction)

Owner: Dos Andes

Winemaker: Louis Vallet

Consulting winemakers: Patrick Piuze, Pascal Marchand

Viticulturist: Carlos Carrasco

Known for: Chardonnay, Pinot Noir

Signature wine: Llai Llai Chardonnay

Visiting: Open to the public by appointment only

Llai Llai's Chardonnay comes from Dos Andes's 618-acre Miraflores estate, located adjacent to the Bío-Bío River, while fruit for the Pinot Noir comes from the Yumbel vineyard, situated just thirty-seven miles from the Pacific Ocean. There's a strong French influence, with Patrick Piuze of Chablis advising on the Chardonnay and Pascal Marchand, formerly winemaker at Burgundy's Domaine de la Vougeraie, consulting on the Pinot Noir. Louis Vallet, a fifth-generation winemaker from Gevrey-Chambertin, oversees the day-to-day winemaking. Llai means "wind" in Mapuche, the language of the indigenous people of Bío-Bío, and when repeated it means "big wind" or "windy."

Loma Larga

Founded: 1999

Address: Fundo Loma Larga sin número, Casablanca, Aconcagua, www.lomalarga .com

Owner: Felipe Díaz Santelices

Winemaker: Cedric Nicolle

Viticulturist: Germán de la Maza

Known for: Syrah, Cabernet Franc, Merlot, Malbec, Pinot Noir, Sauvignon Blanc, Chardonnay

Signature wine: Rapsodia

Other labels: Quinteto

Visiting: Open to the public by appointment only

Loma Larga, meaning "long hill," is a 1,730-acre estate with 363 acres planted to vines. It is unusual for Casablanca in that the winery cultivates more red fruit than white, and it was the first in the valley to plant Cabernet Franc. The super-modern winery is built into the hillside. On the rooftop are planted vines that seamlessly merge into the hill. You can book a helicopter ride over the property to get a good look. In an effort to address regional challenges, including labor and energy costs, Loma Larga has formed an alliance with Miguel Torres and other Chilean winery executives. Under this arrangement, Torres distributes Loma Larga wines nationally, boosting the visibility of Loma Larga and diversifying the Torres offerings.

Los Boldos

Founded: 1991

Address: Camino Los Boldos sin número, Requínoa, Cachapoal, www.clboldos.cl

Owner: Sogrape Vinhos SA

Winemaker: Stephane Geneste

Viticulturist: Juan Pablo Aranda Meyer

Known for: Cabernet Sauvignon, Merlot, Syrah, Chardonnay

Signature wines: Chateau Los Boldos Grand Cru, Amalia

Other labels: Chateau Los Boldos, Momentos Reserva, Sensaciones

Visiting: Open to the public by appointment only; tasting room

The 618-acre Los Boldos sits near the city of Requínoa in the Cachapoal Andes (or Alto Cachapoal, as the region was previously known). Established by a French family, the winery was purchased in 2008 by the Portuguese Sogrape company, which made its fortune in Mateus and also owns Argentina's Finca Flichman. Los Boldos instigated a vast program of reform and investment in the operation, and the emphasis on higher-end wines from the older vines is paying off.

Los Vascos

Founded: 1988

Address: Camino Pumanque, km 5, Peralillo, Colchagua, www.vinalosvascos.com

Owner: Les Domaines Barons de Rothschild (Lafite)

Winemaker: Marcelo Gallardo

Viticulturist: Marcelo Gallardo

Known for: Cabernet Sauvignon

Signature wine: Le Dix de los Vascos

Visiting: Open to the public by appointment only; tasting room

From its 8,900-acre Cañeten property, situated in western Colchagua, not far from Marchigüe and around twenty-five miles from the coast, this French-owned project is one of the first foreign-sponsored winemaking operations in Chile. Like others, it focused at first on trying to make Bordeaux-styled wines in South America, but it has slowly shifted its efforts to making the best Chilean wines it can. It has the largest contiguous vineyard in Colchagua, encompassing more than 1,400 acres of grapes.

Luis Felipe Edwards

Founded: 1976

Address: Vitacura 4130, Santiago, www.lfewines.com

Owner: Luis Felipe Edwards

Winemaker: Nicolas Bizzarri

Consulting winemaker: Matt Thomson

Viticulturist: Eugenio Cox

Known for: Cabernet Sauvignon, Malbec, Syrah, Carmenère, Pinot Noir, Sauvignon Blanc, Chardonnay, Mourvèdre, Grenache, Merlot

Signature wines: Doña Bernarda, LFE900 Blend, LFE900 Malbec

Other labels: Marea de Leyda, Pupilla, Alto los Romeros, Autoritas, Claro, Dancing Flame, Santa Luz, Terra Vega

Visiting: Open to the public by appointment only; tasting room

The history of this very successful winery dates back to 1976, when Luis Felipe Edwards Sr. purchased the Fundo San José de Puquillay estate, located in Colchagua, and planted it with 148 acres of vines from the early 1900s, mainly Cabernet Sauvignon. Today it claims to be the largest 100 percent family-owned wine company in Chile, with over 4,500 acres of estate vineyards. It owns vineyards in Pupilla, western Colchagua, 988 acres in Maule's Retiro, and a 400-acre estate in San Antonio's Leyda. In January 2013 it acquired an additional 1,730 acres bordering the Nilahue estuary in the coastal Colchagua Valley. If it can maintain its standards, this expansion is good news for wine lovers. Matt Thomson, a New Zealander, was recently brought on to assist with white wines and Pinot Noir.

Maquis

Founded: 1927

Address: Benjamín 2965, Las Condes, Santiago, www.maquis.cl

Owner: Hurtado Vicuña family

Winemaker: Juan Alejandro Jofré

Consulting winemakers: Xavier Choné, Jacques Boissenot, Eric Boissenot

Viticulturists: Jaime Gatica, Xavier Choné

Known for: Cabernet Franc, Carmenère, Syrah

Signature wines: Lien, Franco

Other labels: Calcu

Visiting: Open to the public by appointment only

The Hurtado family has owned the 336-acre Viña Maquis vineyard for over a century, but for decades the coveted grapes were sold to other wineries. It wasn't until about ten years ago that they decided to make their own estate wines. The unique vineyard is bordered by the Tinguiririca River and the Chimbarongo Creek, two large waterways that once brought alluvial sediments from the Andes and contribute to the complexity of Maquis' terroir. The winery is easy on the eyes and fits seamlessly into the vineyard setting. The confident but shy Jofré is doing a great job with these wines.

Calcu (www.calcu.cl) is the winery's other label, with wines made from a combination of grapes from young vines on the estate and purchased grapes from other spots in Colchagua. Maquis wines are made exclusively from mature estate fruit. Xavier Choné, who consults for these wines and the estate bottlings, also works with Bordeaux's Château d'Yquem, Leoville Las Cases, and Napa Valley's Opus One and Dalla Valle.

Matetic

Founded: 1999

Address: Fundo Rosario sin número, Lagunillas, Casablanca, Aconcagua, www.matetic.com

Owner: Matetic family

Winemaker: Julio Bastias

Consulting winemaker: Rodrigo Soto

Viticulturist: Alan York

Consulting viticulturist: Pedro Parra

Known for: Syrah, Sauvignon Blanc, Chardonnay, Pinot Noir

Signature wine: Matetic Syrah

Other labels: EQ, Corralillo

Visiting: Open to the public; tasting room; restaurant (Equilibrio), lodge (La Casona)

The family owners of this winery have Croatian roots in Chilean Patagonia, where they own 247,000 acres of pastureland for sheep and dairy farming (making them among Chile's largest landholders). The Matetics also made a lot of money through an ironworks foundry. Twenty years ago they bought the massive 40,000-acre property in the Rosario Valley, on the border between Casablanca and San Antonio, of which 300 acres are farmed biodynamically. A winery that started strong and then lost its way, it is now back on track. The restaurant and lodge are memorable.

Maycas del Limarí

Founded: 2005

Address: Nueva Aurora, Ovalle, Limarí, www.maycasdellimari.com

Owner: Concha y Toro Group

Winemakers: Javier Villarroel, Marcelo Papa, Francesca Perazzo

Known for: Chardonnay, Pinot Noir, Syrah

Signature wine: Quebrada Seca Chardonnay

Visiting: Open to the public by appointment only

In the Quechua language, Maycas means "arable lands," which, together with Inti ("sun"), were the pillars of the Inca culture.

The wine labels pay tribute to the famous Inca solar calendar, used to determine agricultural cycles. A rising star among Chilean wineries, Maycas del Limarí produces stunning, site-specific wines under the direction of Marcelo Papa, Concha y Toro's chief winemaker, from a special plot containing vast amounts of quartz and limestone located in the Quebrada Seca area. Peter Richards notes in his online "Chile Wine Brief" (2013) that the winery will be introducing a *méthode traditionnelle* sparkling wine made from Pinot Noir and Chardonnay in the near future.

Mayu. *See* Falernia

Meli

Founded: 2002

Address: Fundo El Peumal sin número, San Javier, Maule, www.meli.cl

Owner: Adriana Cerda

Winemaker: Adriana Cerda

Viticulturist: Adriana Cerda

Known for: Carignan

Signature wine: Dueño de la Luna

Visiting: Open to the public by appointment only

Meli is the culmination of a dream of Adriana Cerda, a respected winemaker with more than thirty years' experience. Her résumé includes a long tenure at De Martino, where she oversaw the launch of Carmenère in 1996. She and her three adult sons bought a thirty-two-acre, dry-farmed property with sixty-year-old Carignan and Riesling vines. *Meli* means "four" in the Mapuche language: the name represents the four family members. The label artwork is a take on the Andean cross, which is found in various forms among the indigenous peoples of Latin America. Meli is a member of MOVI.

Miguel Torres

Founded: 1979

Address: Longitudinal Sur, km 195, Casilla 163, Curicó, www.migueltorres.cl

Owner: Miguel Torres

Winemaker: Miguel Torres

Viticulturist: Miguel Torres

Known for: Cabernet Sauvignon, Carmenère, Carignan, Syrah, Chardonnay, Sauvignon Blanc

Signature wine: Conde de Superunda

Other labels: Manso de Velasco, Cordillera, Santa Digna, Las Mulas, Hemisferio

Visiting: Open to the public; tasting room; restaurant

Torres was the first foreign winery to invest in Chile in the modern era, and it has had a transformative effect on Chilean winemaking. From winery hygiene and care to temperature-controlled winemaking to using modern presses, Torres has led the way. Since the mid-1980s, it has also been heavily involved in clonal selection and importing plant material. Torres's 911-acre property near Empedrado, in coastal Maule, is touted by Peter Richards in *Wines of Chile* as "perhaps the first vineyard in Chile to be planted primarily on the basis of soil type—and probably the first vineyard in Chile to be planted on slate." Slate is a distinctive characteristic of the vineyards in Spain's Priorat and Germany's Mosel. Torres's signature wine, "Conde de Superunda," is a superb proprietary blend of Tempranillo, Cabernet Sauvignon, Monastrell, and Carmenère. Torres is also dedicated to rediscovering Chilean "heritage" wines, based on traditional Pais, in table red (Reserva de Pueblo) and sparkling (Estelado) bottlings. The winery is visited annually by thousands of local and foreign tourists.

Misiones de Rengo

Founded: 2001

Address: Daniel Morán sin número, Rengo, Cachapoal, www.misionesderengo.cl

Owner: VSPT Wine Group

Winemaker: Sebastián Ruiz Flaño

Viticulturalist: Rodrigo Silva

Known for: Carmenère, Cabernet Sauvignon, Sauvignon Blanc

Signature wine: Gran Reserva Cuvée Carmenère

Visiting: Open to the public; tasting room

Misiones de Rengo Vineyard pays tribute to the Spanish missionaries who brought the first vines to Chile and settled in a small Rapel Valley town, named Rengo in memory of a brave Indian chief. Ruiz Flaño has experience in both France's Bordeaux and California's Napa Valley and has been with the winery since its inception. Since 2006, the winery has been among the most popular in Chile. In 2011 it commanded a 10.6 percent market share in the very competitive range of bottles priced at 1,200 pesos or above.

Montes Wines. *See* Viña Montes

MontGras

Founded: 1993

Address: Camino Isla de Yáquil sin
número, Palmilla, Colchagua,
www.montgras.cl

Owners: Eduardo Gras, Hernán Gras,
Cristián Hartwig

Winemaker: Santiago Margozzini

Viticulturist: Ricardo Araneda

Known for: Carmenère, Syrah, Cabernet
Sauvignon, Sauvignon Blanc

Signature wine: Carmenère Reserva

Visiting: Open to the public; tasting room

From its original start in Palmilla with
Hernán Gras's winemaking skill and the
business sense of his brother, Eduardo,
and Cristián Hartwig, MontGras has rapidly expanded and has now acquired more
land in Colchagua, including 988 acres
in Pumanque. A pioneer from the start,
Viña MontGras was among the first wineries in Latin America to introduce large-
scale lyre trellising of vines. It has also
been a leader in adopting new technology and, along with five other local wineries, helped establish the Colchagua Valley
Wine Route to promote tourism and educate the public about the traditions of Chilean winemaking.

MontGras Properties (San José
Properties)

See MontGras, Ninquén

When Hernán Gras, his brother, Eduardo, and their business partner, Cristián Hartwig, all from veteran wine families, decided to join forces and form San
José Properties (SJP), nobody (except perhaps the three of them) anticipated that
they would ascend the winemaking lad-
der in Chile as quickly as they did. They
have been expanding fast, and their empire now includes Intriga, based on 455
acres in Maipo, and the impressive Ama-
ral, based in Leyda on 1,400 acres and focusing exclusively on white wines.

Morandé

Founded: 1996

Address: Rosario Norte 615, Oficina 2101,
Las Condes, Santiago, www.morande.cl

Owner: Grupo Belén

Winemakers: Ricardo Baettig, Pablo
Morandé

Viticulturist: Diego Correa

Known for: Carignan

Signature wine: House of Morandé

Other labels: Pionero, Vigno

Visiting: Open to the public; tasting room;
store (House Casa del Vino); restaurant

Pablo Morandé, the winery's founder, is the
acknowledged godfather of Casablanca, the
one who put it on the Chilean wine map.
He sold the estate to Grupo Belén in 2000
but continues to serve as technical director.
The new owners have helped to focus the
portfolio and provided capital for extensive
new vineyard projects, such as the acquisition of 321 acres in Lo Ovalle, which Morandé has planted to aromatic whites and
Pinot Noir. The House Casa del Vino, in
Casablanca, is a showcase for all of Belén's
wines, including Morandé, Vistamar, Man-
cura, Tiraziš, and their Argentinean holdings, Zorzal and Passionate Wine. The facility also has a great restaurant, picnic
area, and children's play space.

Nativa Eco Wines

Founded: 2009

Address: Av. Apoquindo 3669, Piso 7, Las Condes, Santiago, www.nativawines.com

Owner: Fundación Claro Vial

Winemaker: Felipe Ramírez

Viticulturist: Sebastián Warnier

Known for: Cabernet Sauvignon, Sauvignon Blanc, Chardonnay, Carmenère, Gewürztraminer

Signature wine: Gran Reserva Cabernet Sauvignon

Visiting: Open to the public by appointment only

This winery began life in 1995 as part of the Carmen estate. Its wines were among the first in Chile to be made with certified organic grapes. In 2007, Nativa began to work closely with organic-grape producers from Maipo Valley and in 2008 brought in the first organic grapes from its own vineyards in Marchigüe in the Colchagua Valley. A year later, a new winery was created with a commitment to organics. Felipe Ramírez has been instrumental in improving wine quality, especially the Carmenère and Cabernet Sauvignon.

Neyen

Founded: 2002

Address: Apalta Road, km 11, Colchagua, www.neyen.cl

Owner: Huneeus family

Winemaker: Rodrigo Soto

Viticulturist: José Aguirre

Known for: Carmenère, Cabernet Sauvignon

Signature wine: Espíritu de Apalta

Visiting: Open to the public by appointment only; tasting room

The Neyen winery was founded on the site of one of Apalta's first wineries, which dates back to 1890. The Rojas family, who purchased the property in 1973, knew that the 3,300-acre property they had owned for more than thirty years was unique, and they created Neyen to showcase its remarkable vineyard. In 2012, they sold the property to the Huneeus family, although they remain active partners in the project. Along with Montes and Lapostolle, the Rojas family helped bring the illustrious Apalta region to prominence in Chile.

Ninquén

Founded: 1998

Address: Av. Eliodoro Yáñez 2962, Providencia, Santiago, www.ninquen.cl

Owners: Eduardo Gras, Hernán Gras, Cristián Hartwig

Winemaker: Santiago Margozzini

Viticulturist: Ricardo Araneda

Known for: Syrah, Cabernet Sauvignon, Carmenère

Signature wine: Antu Syrah

Visiting: Open to the public; tasting room

Developed by MontGras in the late nineties, Ninquén, which means "plateau on a mountain" in Mapuche, is a three-wine effort based on a namesake hillside vineyard in Colchagua. The eponymous wine is a *corte* with Cabernet Sauvignon and Syrah, while their two other reds, sold under the less expensive Antu label, include a pure Syrah and a blend of Cabernet

Sauvignon and Carmenère. *Antu* means "sun" in Mapuche.

Ochagavía

Founded: 1851

Address: Av. Providencia 929, Providencia, Santiago, www.ochagaviawines.com

Owner: Watt's SA

Winemaker: Andrés Caballero, Iván Martinovic

Consulting winemaker: Nick Goldschmidt

Viticulturist: Samuel Barros

Known for: Carmenère, Cabernet Sauvignon, Sauvignon Blanc, Chardonnay, Merlot

Signature wine: Raíces Nobles

Other labels: 1851, Silvestre, Media Luna, Espuela

Visiting: Open to the public by appointment only

Silvestre Ochagavía, the winery's founder, is one of the founding fathers of Chilean wine. In the nineteenth century, he was among the first to emphasize French grapes. Today the winery, owned by Carolina Wine Brands, produces a range of wines primarily sourced from the Maipo Valley.

OchoTierras

Founded: 2006

Address: Parcel 44 B, Colonia Limarí, Ovalle, Coquimbo, www.ochotierras.cl

Owners: Ronald Cuellar, Horacio Alvarez, Rodrigo Guerrero

Winemaker: Rodrigo Guerrero

Consulting winemaker: Dominic Hentall

Viticulturist: Rodrigo Guerrero

Known for: Syrah, Cabernet Sauvignon, Carmenère

Signature wines: Gran Reserva, Reserva Single Vineyard

Visiting: Open to the public by appointment only

This young winery is the initiative of three local agriculturists who, notwithstanding their youth, have considerable experience in growing wine grapes for fine-wine production in the Limarí Valley. They farm 173 acres of grapes, planted on different parcels. The Carmenère, Cabernet Sauvignon, and Sauvignon Blanc vineyards are adjacent to the winery in the town of Limarí, while the Syrah vineyards are located farther into the eastern *cordillera* (mountain range), and the Chardonnay vineyards are closer to the coast, in Cerrillos de Tamaya.

Odfjell

Founded: 1997

Address: Camino Viejo a Valparaiso 7000, Padre Hurtado, Santiago, www.odfjellvineyards.cl

Owner: Odfjell family

Winemaker: Arnaud Hereu

Consulting winemaker: Patrick Ducournau

Viticulturist: Arturo Labbé

Known for: Cabernet Sauvignon, Carignan, Malbec, Carmenère, Syrah

Signature wine: Orzada

Visiting: Open to the public by appointment only; tasting room

The winery is owned by the family of the Norwegian shipping magnate Dan Odfjell, who arrived in Chile on business and ended up buying land in the hills west of Santiago in central Maipo. The land was initially planted to orchards, but in 1992 the conversion to vineyards began, with the first wines released seven years later. The gravity-flow winery, designed by architect Laurence Odfjell, the son of the founder, is situated on a slope above the vineyards, with 60 percent of the structure actually built into the slope and underground. Patrick Ducournau, who invented the micro-oxygenation process in Madiran, France, is Odfjell's consulting winemaker. It is a member of Vigno, Maule's old-vine Carignan consortium.

A Catalan Frenchman who settled in Chile over ten years ago, Arnaud Hereu works full time as winemaker at Odfjell, but he has also recently initiated a postage-stamp-sized project called Hereu (www.hereu.cl), which is a member of MOVI. This project produces two wines: a Chardonnay and his namesake red *corte,* which combines Syrah, Carignan, and Malbec. Hereu means "the heir who will inherit the estate." Arnaud's children continue in the tradition as the designers of his labels.

O. Fournier Chile

Founded: 2006

Address: Camino a Constitución, km 20, San Javier, Maule, www.ofournier.com

Owner: O. Fournier Group

Winemaker: José M. Spisso

Viticulturist: Hugo Donoso

Known for: Cabernet Franc, Carignan, Sauvignon Blanc

Signature wine: Alfa Centauri Red Blend

Other labels: Centauri, Urban Maule

Visiting: Open to the public by appointment only; tasting room and restaurant; guesthouse

With additional and quite successful winery operations in Argentina and Spain, José Manuel Ortega Gil-Fournier has become one of the world's most admired wine personalities. His Chilean adventure took him on three years of exploration before he decided on the right locations for his red wines (a dry-farmed parcel in the Maule Valley with vines 65 to 120 years old) and white wines (eighty-seven acres in Lo Abarca, San Antonio). A former banker for Goldman Sachs in New York and Santander in Madrid, Ortega owns the only multinational winery group in which wines from each of its wineries has been named the country's best wine (Alfa Crux 2001 in Argentina, Spiga 2004 in Spain, and O. Fournier Centauri Sauvignon Blanc 2007 in Chile).

Pérez Cruz

Founded: 2002

Address: Fundo Liguai sin número, Huelquen, Paine, Maipo, www.perezcruz.com

Owner: Pérez Cruz family

Winemaker: Germán Lyon

Consulting winemaker: Álvaro Espinoza

Viticulturist: Gonzalo Prat

Known for: Cabernet Sauvignon

Signature wines: Cabernet Sauvignon Reserva, Viña Pérez Cruz

Other labels: Quelen, Chaski

Visiting: Open to the public by appointment only; tasting room

Acquired by the Pérez family in 1963, the 1,300-acre estate was initially developed for cattle and other crops, including alfalfa and almonds. The first 346 acres of vines were planted in two stages in the mid- to late 1990s. Although they originally intended only to grow and sell grapes, over time the family decided to build a brand. They began construction of their state-of-the-art winery, designed by José Cruz Ovalle, in the early 2000s. Pérez Cruz focuses on red wines, with Cabernet Sauvignon accounting for over three-quarters of its production. The estate has 642 acres of vineyard at elevations between 1,450 and 1,650 feet.

Polkura

Founded: 2002
Address: Chequen sin número, Marchigüe, Colchagua, www.polkura.cl
Owners: Muñoz family, Bruchfeld family
Winemaker: Sven Bruchfeld
Viticulturist: Sven Bruchfeld
Known for: Syrah
Signature wine: Block G+I
Visiting: Open to the public by appointment only

Formerly the winemaker at Santa Carolina, Sven Bruchfeld is affable, passionate, straightforward, and opinionated. He and his partner, Gonzalo Muñoz (whom he met in France), created this small MOVI-member winery. They focused on Syrah and selected Marchigüe as the right spot for this variety. *Polkura,* which means "yellow stone" in Mapuche, is the name of a lit-

tle hill located inside the thirty-acre vineyard. It also refers to the large amounts of yellow granite in the clay soils.

Porta

Founded: 1954
Address: Camino a la Compañía, sin numéro, La Morera, Codegua, Cachapoal, www.portawinery.cl
Owner: Dos Andes
Winemaker: Ana Salomó
Viticulturist: Carlos Carrasco
Known for: Carmenère, Cabernet Sauvignon, Sauvignon Blanc, Chardonnay, Merlot, Pinot Noir, Syrah
Signature wine: Cima
Other labels: Organic "Boldo"
Visiting: Open to the public by appointment only

Ana Salomó started in India, of all places, as an assistant for the very good *méthode traditionnelle* producer Omar Khayyam (now closed). She gained experience across a swath of Chile's notable wineries—Santa Rita, Miguel Torres, and Santa Carolina—before landing at Quebrada de Macul, where she collaborated with Ignacio Recabarren, one of Chile's most respected and renowned winemakers. At Porta she oversees a wide range of wines in an elegant colonial structure that also houses a remarkable six-room guesthouse.

Quebrada de Macúl

Founded: 1996
Address: Consistorial 5900, Peñalolen, Santiago, www.domusaurea.cl
Owners: Ricardo and Isabel Peña

Winemaker: Jean-Pascal Lacaze

Consulting winemaker: Patrick Valette

Viticulturist: Jean-Pascal Lacaze

Known for: Cabernet Sauvignon

Signature wine: Domus Aurea

Other labels: Alba de Domus, Stella Aurea, Peñalolén, Pargua

Visiting: Open to the public by appointment only

Quebrada de Macúl is a family-owned project based on a 45-acre hillside vineyard planted to Bordelaise red grapes back in 1970. Back then it was not common to plant on slopes or at elevation. Patrick Valette and Jean-Pascal Lacaze joined the operation in 2003. The winery is celebrated for its low-yield, high-concentration efforts. Domus Aurea consistently rates as one of Chile's iconic reds.

Quintay

Founded: 2005

Address: Ruta 68, km 62.5, Casablanca, Aconcagua, www.quintay.com

Owners: Consortium of 12 owners

Winemaker: Vicente Johnson

Viticulturist: Vicente Johnson

Known for: Sauvignon Blanc, Pinot Noir, Syrah

Signature wine: Quintay Q

Other labels: Clava

Visiting: Open to the public; tasting room

The name Quintay comes from the native Mapudungun language; it means "go with the flow." The winery's symbol, shown on its labels, is a *clava,* the carved and polished stone emblem used by the local Picunche tribal chiefs. Quintay works stra-tegically with the well-regarded Álvaro Espinoza's GEO consulting team and eight different growers throughout the Casablanca region and San Antonio. It specializes in Sauvignon Blanc.

Rukumilla

Founded: 2000

Address: Camino El Mazano 021, Lonquén, Talagante, Maipo, www.rukumilla.com

Owner: Pedro Andrés Costa Lagos

Winemaker: Pedro Andrés Costa Lagos

Viticulturist: Pedro Andrés Costa Lagos

Known for: Cabernet Sauvignon, Malbec, Cabernet Franc, Syrah

Signature wine: Rukumilla

Visiting: Open to the public by appointment only

The minuscule Rukumilla winery, a member of MOVI, is a labor of love of the passionate Pedro Andrés Costa and his wife, Angelica Grove, who work from a five-acre organic parcel adjacent to their home. Rukumilla, a Mapuche word meaning "golden breasts," is the name of a woman who lived in Araucanía and was the daughter of a Millantu chief and a foreign captive.

San José de Apalta

Founded: 1970

Address: Longitudinal Sur, km 107, Rosario, Rengo, Cachapoal, www.sanjosedeapalta.cl

Owner: Agrícola Santa Cristina Ltda.

Winemaker: Raimundo Barros

Consulting winemaker: Stephan Geneste

Viticulturist: José Ramón Vega Orueta

Known for: Cabernet Sauvingon, Carmenère, Syrah, Chardonnay

Signature wine: Friends Collection

Visiting: Open to the public by appointment only; tasting room

The winery sources fruit from three different areas—Peumo, in the heart of Cachapoal; Rosario, which is slightly south of Rancagua in the Alto Cachapoal; and Rengo, at the winery estate located slightly south of Rosario. This is another winery whose owners started out selling grapes before taking the winery plunge in the mid-1990s. Though young, San José de Apalta has enjoyed success.

San Pedro

Founded: 1865

Address: Av. Vitacura 4380, Piso 7, Vitacura, Santiago, www.sanpedro.cl

Owner: VSPT Wine Group

Winemakers: Marco Puyo, Marcelo García, Gonzalo Castro, Miguel Rencoret, Carlos Chandia

Viticulturist: Raul Aquinas Wunkhaus Hamdorf

Known for: Cabernet Sauvignon, Syrah, Carmenère, Sauvignon Blanc

Signature wine: Cabo de Hornos

Other labels: 1865, 35° South B.O., Gato Negro, Castillo de Molina, Kankana del Elqui, Tierras Moradas, Epica

Visiting: Open to the public by appointment only; tasting room

This long-established winery and its main vineyards are located in Molina, 125 miles south of Santiago. It holds close to 3,000 acres that form one of the longest stretches of continuous vineyards in Latin America. San Pedro has another 3,700 acres of vineyards in the Central Valley, spread among Elquí, Casablanca, San Antonio–Leyda, Maipo, Cachapoal, Maule, and Bío-Bío. In 1994, the winery went public and was acquired by the Chilean brewing giant CCU, part of the corporate empire of the wealthy Luksic family. "Cabo de Hornos" is San Pedro's top wine. The "1865" line comprises red varietal wines, each sourced from a different region.

The winery's tourism offerings range from standard tours to visits with lunch or a picnic to horseback riding. From time to time they offer a special tour that visits the VSPT group's wineries in three different valleys: San Pedro (in Curicó), Santa Helena (in Colchagua), and Altaïr (in Cachapoal).

Santa Alicia

Founded: 1954

Address: Circunvalación Máximo Valdés 4135, Pirque, Santiago, www.santa-alicia.cl

Owner: Bethia Group

Winemakers: Eduardo Gajardo, David González

Viticulturist: Fernando Becerra

Known for: Cabernet Sauvignon, Merlot, Carmenère, Chardonnay, Sauvignon Blanc, Cabernet Franc, Petit Verdot, Malbec, Syrah

Signature wine: Millantu

Other labels: Anke, Gran Reserva de Los Andes

Visiting: Open to the public by appointment only; tasting room

The winery was founded under the name Casas de Pirque by the agricultural secretary of the time, Máximo Valdés Fontecilla. Under its new name and new ownership, Viña Santa Alicia exports most of its wines, which it sources from its 494 acres of vineyards in Pirque, Melipilla, and María Pinto. Santa Alicia (1204–50) is the patron saint of the blind and paralyzed.

Santa Carolina

Founded: 1875

Address: Til Til 2228, Macul, Santiago, www.santacarolina.cl

Owner: Watt's SA

Winemaker: Andrés Caballero

Consulting winemaker: Nick Goldschmidt

Viticulturist: Samuel Barros

Known for: Carmenère, Cabernet Sauvignon, Sauvignon Blanc, Chardonnay

Signature wine: Herencia

Other labels: VSC

Visiting: Open to the public by appointment only

In 1875, Luis Pereira Cotapos, a Chilean businessman, planted French Bordelaise vines on his property and had the insight to invite a winemaker or two from France to come make his wines. He named the estate Santa Carolina, after his wife. Years later, the company, now Carolina Wine Brands, has evolved into one of the most significant in the country. The original warehouse, declared a national monument in 1973, is noteworthy for, among other things, being one of the few still standing that was constructed using an ancient mortar of lime and egg whites. Now holding vineyards all across Chile's premier wine zones, Caro-

lina Wine Brands is a fixture of the wine landscape, making a range of wines from the popular to the iconic.

Santa Ema

Founded: 1931

Address: Izaga 1096, Casilla 17, Isla de Maipo, Maipo, www.santaema.cl

Owner: Pavone family

Winemakers: Andrés Sanhueza (chief winemaker), Claudio Pardo, Rodrigo Blazquez

Consulting winemaker: Irene Paiva

Known for: Carmenère, Merlot, Cabernet Sauvignon, Chardonnay, Sauvignon Blanc

Signature wines: Santa Ema Reserve Merlot, Santa Ema Catalina

Other labels: Rivalta, Amplus

Visiting: Open to the public by appointment until tasting salon opens

Santa Ema was established by Pedro Pavone, the son of a Piemontese winemaker who emigrated from Italy in 1917. Pavone planted several hundred acres of vines in Peumo's Rosario area in 1931, though the winery was not established until the 1950s. On the initiative of his son, Felix, the winery began exporting in the 1950s and remains one of Chile's more successful export wineries. The winery is advised by the extremely talented Irene Paiva, formerly of Caliterra and San Pedro and a founding director of MOVI. Santa Ema works from a very modern and sustainably operated high-tech winery in El Peral, established in 2003, and its reputation for good-value wines is contributing to its growing success.

Santa Helena

Founded: 1942

Address: Av. Vitacura 4380, Pisos 6 & 7, Vitacura, Santiago, www.santahelena.cl

Owner: VSPT Wine Group

Winemakers: Matías Rivera, Maite Hojas

Viticulturist: Marcelo Lorca

Known for: Cabernet Sauvignon, Syrah, Carmenère, Sauvignon Blanc

Signature wine: Don

Other labels: Notas de Guarda, Parras Viejas, Vernus, Selección del Directorio

Visiting: Open to the public by appointment only

Santa Helena was acquired by the Viña San Pedro Group in 1994. With a total of 825 acres in Colchagua, spread over three properties, and long-term contracts in Maipo, Elquí, Casablanca, Leyda, Cachapoal, Curicó, and Maule, it is a formidable operation. Since the early 2000s, the winery has been given a long leash of independence from the parent company and has been flexing its muscles. It accounts for about 10 percent of the turnover of the VSPT group.

Santa Rita

Founded: 1880

Address: Camino Padre Hurtado 0695, Alto Jahuel, Santiago, www.santarita .com

Owner: Fundación Claro Vial

Winemakers: Andrés Ilabaca, Cecilia Torres, Carlos Gatica

Consulting winemaker: Brian Croser

Viticulturist: Sebastián Warnier

Consulting viticulturist: Eric Boissenot

Known for: Cabernet Sauvignon, Sauvignon Blanc, Chardonnay, Carmenère, Merlot, Syrah

Signature wine: Casa Real

Other labels: Medalla Real, Pehuén, Triple C, Floresta, 120

Visiting: Open to the public; tasting room; restaurant (Doña Paula, Café La Panadería); hotel (Casa Real)

In 2005, the Santa Rita Group's commercial structure was overhauled, and the company acquired 2,965 acres of new land in coastal Colchagua (Pumanque) and 3,660 in coastal Limarí, near Tabalí—and added to its existing holdings in Casablanca, Maipo, and Palmilla in Colchagua. Over time, the namesake winery has consolidated its position as one of Chile's most successful and innovative estates. Its range begins with the highly successful "120" series of wines, named in honor of 120 patriots who helped lead Chile to independence in the early nineteenth century. With a broad range of offerings, the winery is acclaimed for its ultra-premium and highly acclaimed Casa Real and Triple C wines.

Santa Rita Group

See Carmen, Nativa Eco Wines, Santa Rita

The massive Santa Rita Group vies with Carolina Wine Brands for the position of Chile's second-largest wine company, just behind the Concha y Toro Group. Established in the late nineteenth century at its present site in Alto Jahuel, the namesake winery was acquired by the prominent Chilean businessman Ricardo Claro, who also acquired Carmen (1987), Sur Andino (2001), and a large share in Los Vascos

(1996), in addition to hopping the Andes and launching Doña Paula in Argentina in 1998. The more recent brand additions include Viña Centenaria, an offering of wines from both Chile and Argentina.

Seña

Founded: 1995

Address: Av. Nueva Tajamar 481, Torre Sur, Oficina 503, Las Condes, Santiago, www.sena.cl

Owner: Eduardo Chadwick

Winemaker: Francisco Baettig

Viticulturist: Jorge Figueroa

Known for: Bordeaux blend with Carmenère

Signature wine: Seña

Visiting: Open to the public by appointment only; tasting room

Eduardo Chadwick pioneered a joint venture with Robert Mondavi to hand craft a world-class Chilean wine. They took several years to find the best Aconcagua spot and make trial blends. The first wine was released in 1997. The 2001 vintage came in second at the Berlin Tasting in 2004, ahead of Château Lafite Rothschild and other noteworthy French and Tuscan wines. Seña also participated in an Asian tour, presenting a vertical tasting of the wine in several cities. In 2003, when Constellation bought Mondavi, Eduardo had the right of first refusal on all the joint ventures and bought Seña back.

Tabalí

Founded: 1993

Address: Calle La Vina Tabalí, Ovalle, Coquimbo, www.tabali.com

Owners: Luksic family

Winemakers: Felipe Müller, Christián Sepúlveda

Viticulturist: Héctor Rojas

Known for: Syrah

Signature wine: Reserva Syrah

Visiting: Open to the public; tasting room

Tabalí is a joint venture between the giant San Pedro winery and Agrícola y Ganadera Río Negro, the company that originally planted the vineyard in 1993 and was owned by Guillermo Luksic, who sat on the board of San Pedro and also owned Leyda in the San Antonio Valley. He died in March 2013, and the winery is now run by his son, Nicolás Luksic Puga. Most of the vines were planted between 1993 and 1998, making them among the region's earliest, and the operation benefited from the input of the French consultant Jacques Lurton, who was consulting for San Pedro at the time. The winery owns a newer 161-acre vineyard in Talinay, about sixteen miles inland, where there are substantial outcroppings of limestone. Tabalí plans to launch its first sparkling wines in 2014.

Tamaya

Founded: 1997

Address: Camino Quebrada Seca, km 9, Ovalle, Coquimbo, www.tamayawines.cl

Owner: René Merino

Winemaker: José Pablo Martín

Consulting winemaker: Felipe de Solminihac

Viticulturist: Lisardo Álvarez

Known for: Chardonnay, Syrah

Signature wine: T Limited Release line

Other labels: Sweet Goat, Winemaker Gran
Reserva

Visiting: Open to the public by appointment only

Tamaya, which means "high outlook" in the local native language, began as a fruit farm that diversified into viticulture on the advice of the former winemaker Carlos Andrade, who also owned the property opposite Tamaya. It is located in Limarí's cool maritime strip north of the river. Along with Tabalí, it is considered a leader in the area. The 9,900-acre property contains 2,250 acres planted with fruit (cherimoya, papaya, citrus, and pomegranates) and olive groves, and 395 acres planted to vineyards. Consulting enologist Felipe de Solminihac is one of four partners in Viña Aquitania.

TerraMater

Founded: 1996

Address: Luis Thayer Ojeda 236, Piso 6,
Providencia, Santiago, www.terramater
.cl

Owners: Edda Canepa, Antonieta Canepa,
Gilda Canepa

Winemaker: Paula Cifuentes

Consulting winemaker: Álvaro Espinoza

Viticulturist: Andrés Urrutia

Known for: Cabernet Sauvignon, Zinfandel, Sangiovese, Carmenère

Signature wine: Mater

Other labels: Unusual, Altum,
Paso del Sol

Visiting: Open to the public; tasting room

After the split-up of the Canepa winery, the Canepa sisters decided that retirement wasn't the plan. As the family has been in the wine business since the 1930s, they decided to keep much of the family land and have a go at it. They work from established vineyards in Maule, Curicó, and Maipo, many with vines more than fifty years old. They also make magnificent oil (certified organic) from their 618 acres of olive groves and grow various types of fruit. TerraMater is one of the few Chilean producers of very good wines from Zinfandel and Sangiovese.

TerraNoble

Founded: 1993

Address: Fundo Santa Camila sin
número, San Clemente, Talca, Maule,
www.terranoble.cl

Owner: Wolf Von Appen

Winemaker: Ignacio Conca

Consulting winemakers: Felipe de
Solminihac, Domenique Delteil

Viticulturist: Pedro Vega, Yerko
Moreno

Known for: Cabernet Sauvignon,
Carmenère, Merlot, Syrah, Pinot Noir,
Chardonnay, Sauvignon Blanc, Riesling

Signature wines: CA2 Costa Carmenère,
Lahuen Carmenère–Cabernet
Sauvignon

Visiting: Open to the public by appointment only; tasting room

The TerraNoble undertaking is impressive. With some 890 acres across Maule, Casablanca, and Colchagua, it produces high-quality wines with a remarkable precision of style that exemplify the unique characteristics of each variety. The winery is among the country's most critically acclaimed, and its wines are all very well made.

Terraustral

Founded: 1994
Address: Hernando de Aguirre 1915,
    Providencia, Santiago,
    www.terraustral.cl
Owner: Crispi family
Winemaker: Christian Merino
Consulting winemaker: Duncan Killiner
Known for: Carmenère
Signature wine: Car Men Air
Visiting: Open to the public by appointment only; tasting room

A company focusing on everyday wines, Terraustral is most familiar to Americans as the maker of "Car Men Air," a Maule-based (you guessed it) Carmenère sold by Trader Joe's. Additionally, Terraustral makes a range of moderately priced and easy-drinking wines under the PKNT label. Duncan Killiner is a fixture in South America, bringing global expertise to projects on both sides of the Andes, including Uruguay.

Torreón de Paredes

Founded: 1979
Address: Fundo Santa Teresa sin
    número, Rengo, Cachapoal,
    www.torreondeparedes.cl
Owners: Álvaro Paredes, Javier Paredes
    Legrand

Winemaker: Álvaro Paredes
Consulting winemaker: Eugenia Díaz
Viticulturist: Álvaro Paredes
Known for: Sauvignon Blanc, Chardonnay,
    Gewürztraminer, Cabernet Sauvignon,
    Merlot, Syrah, Carmenère, Pinot Noir
Signature wine: Don Amado
Other labels: Valle de Rengo, Valdemoro
Visiting: Open to the public by appointment only; tasting room

Torreón de Paredes is a small-scale family wine producer located east of Rengo. Started by the late Amado Paredes when he was in his seventies, it is now run by two of his seven sons. Known historically for making "correct wines," the company has been investing heavily in improvement across the board, from vineyard to winery.

Tres Palacios

Founded: 1993
Address: Magnere 1543, Providencia,
    Santiago, www.vinatrespalacios.cl
Owner: Patricio Palacios
Winemaker: Camilo Rahmer
Consulting winemaker: Irene Paiva
Viticulturist: Juan Saa
Known for: Cabernet, Merlot
Signature wine: Cholqui
Other labels: Family Vintage
Visiting: Open to the public; tasting room

A family-operated estate winery, Tres Palacios was the first to plant in the Cholqui area of the Maipo Valley, ten miles from the city of Melipilla. Cholqui, which means "bird's head" in the native language, is a closed valley about six miles south of the Maipo River. The Horcón de Piedra, one of the highest peaks in the Coastal Mountain Range, lies at the eastern end of the valley.

Undurraga

Founded: 1885

Address: Camino Melipilla, km 34,
  Talagante, Maipo, www.undurraga.cl

Owners: Picciotto family, José Yuraszeck

Winemakers: Hernán Amenabar (chief);
  Rafael Urrejola (T.H. line); María del
  Pilar Díaz (Volcanes de Chile)

Consulting winemakers: Álvaro Espinoza,
  Philippe Coulon

Viticulturist: Francisco Valdivieso

Known for: Pinot Noir, Cabernet Sauvignon, Syrah, Carmenère, Carignan,
  Merlot, Chardonnay, Sauvignon Blanc,
  Riesling

Signature wines: T.H. (Terroir Hunter),
  Altazor

Other labels: Founder's Collection, Sibaris,
  Aliwen, Sparkling People

Visiting: Open to the public; tasting
  room

Undurraga is one of Chile's largest and most historic houses, with some 4,950 acres of vines. It has four core estates, two each in Colchagua and in Talagante, where the winery is based. Among the winery's many claims to fame is that it was the first Chilean producer to export to the United States back in 1903. After over a century of involvement in its eponymous winery, the Undurraga family left the business in 2006. In recognition of its decades of hard work, the Wines of Chile Association named Undurraga its winery of the year in November 2012. The newer "T.H." (Terroir Hunter) line, overseen by the brilliantly talented Rafael Urrejola, is especially impressive, and the wines are site-specific in their expression. Keep your eye open for the new bodega launched by Undurraga called Volcanes de Chile, whose wines are overseen by María del Pilar Díaz.

Valdivieso

Founded: 1879

Address: Luz Pereira 1849, Lontué, Curicó,
  www.valdiviesovineyard.com

Owner: Licores Mitjans SA

Winemaker: Brett Jackson

Consulting winemaker: Kym Milne

Viticulturist: Jorge Rojas

Known for: Cabernet Sauvignon, Cabernet
  Franc, Merlot, Carignan

Signature wine: Caballo Loco

Other labels: Eclat

Visiting: Open to the public; tasting room

When Alberto Valdivieso founded Champagne Valdivieso in Macul, his was the first company in South America to make sparkling wine. He never imagined that more than a hundred years later, in the late 1980s, the company would expand into a successful and extensive line of wines from Curicó's Lontué, the La Primavera estate, and Maule. Sparkling wines, made by both the Charmat method and the *méthode traditionnelle,* are still a foundation of the winery's portfolio. The Carignan and single-vineyard wines are especially noteworthy.

Ventisquero

Founded: 1998

Address: Camino La Estrella 401, Oficina 5, Punta de Cortés, Rancagua, Cachapoal, www.ventisquero.com

Owner: Don Gonzalo Vial

Winemaker: Felipe Tosso

Consulting winemaker: John Duval

Viticulturist: Ricardo Gompertz

Known for: Cabernet Sauvignon

Signature wine: Pangea Syrah

Other labels: Vertice, Herú, Root:1, Yali, Ramirana, Enclave

Visiting: Open to the public by appointment only; tasting room

Ventisquero is a relatively new wine company launched by Agrosuper, Chile's leading fresh-food company and distributor, which initially focused on creating its own supermarket labels. By 2006, the company had over 3,700 acres of vineyards across different estates and regions. It has the goal of remaining completely self-sufficient. Felipe Tosso, who considered being both a tennis pro and classical guitarist before going into wine, was brought in to help the winery evolve from producing entry-level wines to focusing on more terroir-driven wines and environmental sustainability. Working with the Australian winemaker John Duval is accelerating that transition in the premium lines. Pangea is a joint venture between Duval and Ventisquero.

Ventolera (formerly Litoral)

Founded: 2008

Address: Fundo San Juan de Huinca sin número, Leyda, Aconcagua, www.ventolerawines.com

Owners: Vicente Izquierdo, Stefano Gandolini

Winemaker: Stefano Gandolini

Viticulturist: Miguel Aguirre

Known for: Pinot Noir, Sauvignon Blanc

Signature wine: Ventolera

Visiting: Open to the public by appointment only

The *ventolera* is a strong sea wind that blows uncompromisingly through the heart of the sparsely populated Leyda Valley, where pastureland and eucalyptus groves are more common than grapes. In 2011 the businessman Vicente Izquierdo, who founded the Litoral winery, entered into a partnership with winemaker Stefano Gandolini to operate the 346-acre Ventolera. Gandolini was given carte blanche, and the results have been noteworthy. The labels, designed by the artist Irineo Nicora (the former right hand of the winemaker Mauro von Siebenthal), are playful, fresh, and summery.

Veramonte

Founded: 1990

Address: Route 68, km 66, Casablanca, Aconcagua, www.veramonte.cl

Owner: Huneeus family

Winemaker: Rodrigo Soto

Consulting winemaker: Paul Hobbs

Viticulturist: José Aguirre

Consulting viticulturist: Pedro Parra

Known for: Sauvignon Blanc, Chardonnay, Pinot Noir, Cabernet Sauvignon

Signature wines: Veramonte, Ritual, Primus

Other labels: Cruz Andina

Visiting: Open to the public; tasting room

Veramonte was founded by the Huneeus family in 1990 in the Casablanca Valley's easternmost area. Agustín Huneeus senior is a legend in the Chilean wine industry, having built Concha y Toro into the country's most important producer during the 1960s. In 1971, with the political climate in Chile unstable, Huneeus left for the United States, where he took over the helm of the beverage giant Seagram Worldwide for a time, as well as running Franciscan, a winery in Napa. He went on to build the Quintessa winery in 1989 and purchased Flowers Winery on the Sonoma coast. Veramonte has more than 1,000 acres of vines planted on the 11,000-acre estate. Its stunningly modern and high-tech facility, designed by one of Chile's foremost architects, Jorge Swinburn, belies the winery's traditional approach, which yields wines with a definite Old World sensibility. The winery was the first to introduce world-class hospitality and visitor experiences in the valley. On its property is a reservoir that provides a sanctuary for over two dozen species of birds.

Via Wines

Founded: 1998

Address: Fundo La Esperanza sin número, San Rafael, Maule, www.viawines.com

Owners: "Group of Wine Lovers"

Winemakers: Camilo Viani (Chilcas brand); Edgard Carter (Oveja Negra brand); Claudio Villouta (Chilensis brand)

Viticulturist: Nelson Zagmutt

Known for: Cabernet Sauvignon, Merlot, Carmenère, Syrah, Cabernet Franc, Petit Verdot, Sauvignon Blanc, Chardonnay, Pinot Noir

Signature wines: Red One (Chilcas brand), The Lost Barrel (Oveja Negra brand), Chilensis Lazuli (Chilensis brand)

Visiting: Open to the public; tasting room; restaurant

Via was established when Jorge Coderch, a veteran of Valdivieso, decided to get involved in another wine project. He pulled together a group of investors, including his brother Juan, the U.S. businessman Richard Huber, and the Chilean television star Mario Kreutzberger (also known as Don Francisco). Via is a sizable and significant operation: although it is based in San Rafael in northern Maule, it has over 2,500 acres of vineyards in areas ranging from Limarí to Bío-Bío. The concept was born when the British wine merchant Simon Farr, of Bibendum, approached Coderch about creating a line of wines for the British market. Each wine line has its own winemaker.

Vik

Founded: 2006

Address: Hacienda Vik, Millahue, Colchagua, www.vik.cl

Owner: Alexander Vik

Winemaker: Cristian Vallejo

Consultant: Patrick Valette

Viticulturists: Patrick Valette, Cristian Vallejo

Known for: Cabernet Sauvignon, Carmenère, Cabernet Franc, Syrah, Merlot

Signature wine: Vik

Visiting: Open to the public; tasting room;
restaurant and hotel (Hacienda Vik)

Viña Vik is an astounding project. Located on a lovely 11,000-acre property in
Millahue, not far from Apalta, it's the vision of Alexander Vik, a celebrated Norwegian businessman and probably one of
the richest people in the world. When he
decided to realize his dream of owning a
vineyard in South America, he asked the
famous Patrick Valette, the noted French-
Chilean wine specialist and consultant,
to find the best place for it. No expense
has been spared. In addition to the winery, Vik operates Hacienda Vik, a sixteen-
suite hotel. It is operated by Vik Retreats,
the managers of Estancia Vik José Ignacio,
a resort in Uruguay. The winery has only
a couple of vintages to date, but the signature wine, Vik, is splendid.

Villard

Founded: 1989

Address: La Concepción 165, Oficina 507,
Providencia, Santiago, www.villard.cl

Owner: Thierry Villard

Winemaker: Anamaría Pacheco

Consulting winemaker: Jean-Charles
Villard

Viticulturists: Anamaría Pacheco, Jean-
Charles Villard

Known for: Pinot Noir

Signature wine: Tanagra

Other labels: Expresión Reserve, Grand
Vin, Equis, El Noble

Visiting: Open to the public by appointment only

Thierry Villard, a MOVI member, was an
early believer in the potential of the Casablanca Valley and the idea of pioneering a
boutique family winery. Originally from
France, though educated partly in the
United Kingdom, he established the business after spending fifteen years working for Orlando in Australia, where he developed the Jacob's Creek brand, among
other things. Although Thierry is still very
active in the vineyards and winery, his
son Jean-Charles (Charlie) is slowly taking over. Villard is a minuscule operation
by Chilean standards, with less than sixty
acres of vineyards, though it purchases additional fruit in Maipo.

Viña Aquitania

Founded: 1990

Address: Av. Consistorial 5090, Peñalolen,
Santiago, www.aquitania.cl

Owners: Paul Pontallier, Bruno Prats,
Ghislain de Montgolfier, Felipe de
Solminihac

Winemakers: Paul Pontallier, Bruno Prats,
Ghislain de Montgolfier, Felipe de
Solminihac

Viticulturist: Angela Jara

Known for: Chardonnay

Signature wine: SoldeSol

Other labels: Lazuli

Visiting: Open to the public by appointment only; tasting room

The wines of Viña Aquitania are a result
of the combined efforts of four wine lumi-

naries: Paul Pontallier, the winemaker of Château Margaux; Bruno Prats, the former owner of Château Cos d'Estournel; Ghislain de Montgolfier, the owner of Bollinger Champagne; and the revered Chilean enologist Felipe de Solminihac. The original three partners (before de Montgolfier) had wanted to invest in Chilean vineyard land since the early 1980s to make Bordeaux-style Cabernets. They settled on a plot of land they found close to the Cousiño Macul estate and replanted from orchards. The "SoldeSol" wines, made from vineyards in the far southern Malleco Valley, have been a stunning addition to the lineup.

Viña Casablanca

Founded: 1992
Address: Rodrigo de Araya 1431, Macul, Santiago, www.casablancawinery.com
Owner: Watt's SA
Winemaker: Ximena Pacheco
Consulting winemaker: Nick Goldschmidt
Viticulturist: Samuel Barros
Known for: Sauvignon Blanc, Chardonnay, Pinot Noir, Syrah
Signature wine: Neblus
Other labels: Nimbus, El Bosque, Cefiro
Visiting: By appointment only

The ebullient Ximena Pacheco is one of Chile's leading female winemakers and spent a stint working with Paul Hobbs in California. Her pedigree is reflected in the quality of the wines, made mostly from the 136-acre Isabel estate vineyard. In keeping with the guiding philosophy of the winery's parent company, Santa Carolina, she makes use of only French varieties—nothing Spanish (including criollas) or Portu-

guese. In 2011 Viña Casablanca and Laroche Chile, a subsidiary of the French producer Domaine Laroche in France's Chablis region, formed a joint venture, essentially merging under the Viña Casablanca brand and consolidating their 371 acres in Casablanca.

Viña Chocalán

Founded: 1998
Address: Parcela número 16, Santa Eugenia de Chocalán, Melipilla, Maipo, www.chocalanwines.com
Owner: Toro Harnecker family
Winemaker: Fernando Espina
Consulting winemaker: María del Pilar González T.
Viticulturist: Andrés Urrutia
Known for: Cabernet Sauvignon, Carmenère, Merlot, Syrah, Cabernet Franc, Malbec, Petit Verdot, Pinot Noir, Viognier (Chocalán, Maipo Valley); Chardonnay, Sauvignon Blanc, Gewürztraminner, Pinot Noir, Riesling (Malvilla, San Antonio Valley)
Signature wine: Blend Gran Reserva
Visiting: Open to the public; tasting room; restaurant

Chocalán, which means "yellow blossoms" in the native language, is a project of a very successful Chilean businessman, Guillermo Toro González. A bottle supplier for the industry, he sought to diversify the family business and produce high-quality wines. Among other tourism activities, the winery offers a very cool chocolate and wine tasting, hosts weddings, and leads biking and horseback tours of the property.

Viña Garcés Silva (Amayna)

Founded: 2002

Address: Fundo San Andrés de Huinca,
Camino Rinconada San Juan, Leyda,
Aconcagua, www.amayna.cl

Owner: José Antonio Garcés Silva

Winemaker: Francisco Ponce Sanhueza

Consulting winemaker: Jean Michel
Novelle

Viticulturist: Ignacio Casali

Known for: Sauvignon Blanc

Signature wine: Amayna Sauvignon Blanc

Visiting: Open to the public by appointment only; tasting room

The well-to-do Garcés Silva family has
long been involved in diversified agriculture, livestock farming, and grape growing, as well as the soft-drink, insurance,
and real-estate sectors. They own this
small modern facility and make their
wines mostly from grapes from the San
Antonio Valley and Casablanca. They sell
off most of their fruit, much of it to Montes, and keep only a few of their preferred
lots for their Amayna wines. The project,
which began in 1999, was the first commercial planting of vines in the Leyda
Valley. Over time, they've improved their
vineyard efforts in consultation with the
viticulturist Pedro Parra.

Viña La Rosa

Founded: 1824

Address: Fundo La Rosa, Ruta H 66-G, km
37, Peumo, Cachapoal, www.larosa.cl

Owner: Sociedad Agricola la Rosa Sofruco
SA (Ossa family)

Winemaker: Gonzalo Carcamo

Viticulturist: Eugenio Maffei

Known for: Carmenère

Signature wine: Don Reca

Other labels: La Palma, & Y, La Capitana,
Ossa

Visiting: Open to the public by appointment only; tasting room; restaurant

Viña La Rosa is one of the oldest wineries
in Chile. During its first century of operation, the family focused on expanding its
fruit business, and today it is among the
largest fruit exporters in the world. Of the
nine thousand acres of vineyards owned
by the Ossa family, some two thousand
are planted to wine grapes. One of their
vineyards, La Palmeria de Cocalán, is also
the site of the only national park in northern Chile. Here you'll find the only collection of native Chilean palm trees, which
can grow over one hundred feet tall and
live to be more than one thousand years
old. Prior to making their own wine, the
Ossa family sold grapes to both Concha y
Toro, of which they owned 25 percent until
1995, and Lapostolle.

Viña Maipo

Founded: 1948

Address: Av. Nueva Tajamar 481, Torre
Sur, Oficina 1001, Las Condes, Santiago,
www.vinamaipo.com

Owner: Concha y Toro Group

Winemaker: Max Weinlaub

Known for: Syrah

Signature wine: Limited Edition
Syrah

Other labels: Protegido, Gran Devoción,
Vitral

Visiting: Open to the public by appointment only

Acquired by Concha y Toro in 1968, this successful estate vinifies a sizable range of wines overseen by Max Weinlaub, a Maipo veteran.

Viña Mar

Founded: 2002

Address: Camino Interior Nuevo Mundo sin número, Ruta 68, km 72, Casablanca, Aconcagua, www.vinamar.cl

Owner: VSPT Wine Group

Winemaker: Mauricio Garrido

Viticulturist: Tomás Rivera

Known for: Sauvignon Blanc, Chardonnay, Pinot Noir

Signature wine: Sparkling wines

Visiting: Open to the public; tasting room; restaurant (Viñamar)

Though located in Casablanca with 687 cultivated acres, Viña Mar makes a range of wines from different regions, including Maipo and Maule. The winery has a strong emphasis on *méthode traditionnelle* sparkling wines, with the capacity to produce more than twenty-five thousand cases. Viña Mar is located on the main road between Santiago and Valparaiso, so it is easy to find, and its eponymous restaurant is committed to offering typical and classic Chilean preparations.

Viña Montes

Founded: 1988

Address: Parrela 15, Millahue de Apalta, Santa Cruz, Colchagua, www.monteswines.com

Owner: Aurelio Montes

Winemaker: Aurelio Montes

Viticulturist: Rodrigo Barría

Known for: Cabernet Sauvignon

Signature wines: Montes Alpha M, Folly, Purple Angel

Other labels: Kaiken, Outer Limits, Napa Angel, Star Angel

Visiting: Open to the public by appointment only; tasting room; restaurant

Montes is the among the most important wineries in Chile today. Its contributions to quality, its exploring and championing of new regions, its uncompromising perennial commitment to quality across all wines and price ranges, and its innovation inside and outside the country are virtually peerless. The impressive winery in Apalta is well worth a visit, and the modest restaurant is lovely. Aurelio Montes pays as much attention to hospitality as he does to his wines, and the winery offers numerous tasting packages and a great opportunity to hike through the celebrated Apalta vineyard.

Viña Requingua

Founded: 1961

Address: Fundo Requingua sin número, Sagrada Familia, Curicó, www.requingua.com

Owners: Santiago Achurra Larrain, Ramón Achurra Larrain, Santiago Achurra Hernández

Winemaker: Benoit Fitte

Consulting winemaker: Chris Markell

Viticulturist: Sebastien Cabaret

Known for: Cabernet Sauvignon, Merlot,

Carmenère, Syrah, Pinot Noir, Chardonnay, Sauvignon Blanc

Signature wine: Toro de Piedra

Other labels: Casa Santiago, Arte Noble, Puerto Viejo, Potro de Piedra

Visiting: Open to the public by appointment only; tasting room

In the early 1960s, Santiago Achurra Hernández acquired 2,460 acres of land and a winery with 98 acres of grapes in the hope of building a successful wine company. Today the winery has 1,236 acres cultivated and another 1,384 under contract in Maule and Colchagua. Originally it focused on bulk wine for the domestic market, but over time it has developed a focus on premium wines.

Viña San Esteban

Founded: 1974

Address: 2178 La Florida, San Esteban, Aconcagua, www.vse.cl

Owner: Vicente family

Winemaker: Horacio Vicente

Viticulturist: Gabriel Moraga

Known for: Cabernet Sauvignon, Carmenère, Syrah, Chardonnay

Signature wine: In Situ Laguna del Inca

Visiting: Open to the public by appointment only; tasting room

The winery is based on two estates, La Florida and Paidahuén, close to the town of San Esteban at the eastern end of the Aconcagua Valley. This area was home to an earlier civilization whose enigmatic rock drawings, reminiscent of spiders' webs, can be seen around the vineyards of Paidahuén, which means "good place." One of these drawings is used on the label of Viña San Esteban's In Situ range. In addition to touring the winery and tasting, visitors can get a guided tour of the rock drawings in the vineyard, tour the property on horseback or mountain bike, and enjoy a *parrilla* (barbecue) on the Paidahuén Hill.

Viña Siegel

Founded: 1980

Address: Fundo San Elias sin número, Palmilla, Colchagua, www.siegelvinos .com

Owner: Alberto Siegel Dauelsberg

Winemaker: Gonzalo Pérez

Consulting winemaker: Didier Debono

Known for: Carmenère

Signature wine: Gran Crucero Limited Edition

Other labels: Gran Crucero, Crucero Reserva

Visiting: Open to the public by appointment only; tasting room

Alberto Siegel Dauelsberg is a third-generation Chilean. His grandfather was an architect who built several important buildings in downtown Santiago, including the Chilean Federal Reserve Bank, and his father was a viticulturist who spent most of his career in charge of Viña San Pedro's vineyards near Molina. Alberto was the founder of Sociedad La Laguna, the most important Chilean wine brokerage, a business that he still runs: there are few Chilean vintners who have not dealt with Alberto Siegel in this capacity. With 1,850 acres and the assistance of Pedro Parra on new land in Los Lingues, the winery makes a range of elegant wines and is very food-conscious. When I was there, I enjoyed an exquisitely paired lunch prepared by the well-known chef Pilar Rodríguez.

Viña Tamm

Founded: 2007

Address: Rinconada sin número, Tinguiririca, Colchagua, www.vinatamm.com

Owner: Tamm Family

Winemaker: Andrés Tamm Plesch

Consulting winemaker: Leonardo Contreras

Known for: Cabernet Sauvignon, Merlot, Chardonnay, Syrah, Carmenère, Pinot Noir

Signature wine: Cabernet Sauvignon Reserva

Other labels: Puente Negro

Visiting: Open to the public by appointment only

With a wine tradition dating back to the thirteenth century in Germany, the Tamm family has long been dedicated to the cultivation of wine in the Colchagua Valley as well. A successful and diverse clan, the Tamms are also involved in other agricultural companies, including a dairy and a fruit brokerage, and they own Salmones Chiloé, a company well regarded for salmon farming. Their vineyard holdings consist of nearly 194 acres in Colchagua, Tinguiririca, and Puente Negro.

Viña Tarapacá

Founded: 1874

Address: Fundo El Rosario de Naltahua sin número, Isla de Maipo, Maipo, www.tarapaca.cl

Owner: VSPT Wine Group

Winemakers: Edward Flaherty, Cristián Molina

Viticulturist: Tomás Rivera

Known for: Cabernet Sauvignon, Syrah, Carmenère, Sauvignon Blanc

Signature wine: Tara.Pakay

Other labels: Zavala, Gran Reserva Etiqueta Negra, Tarapacá +Plus, Terroir

Visiting: Open to the public by appointment; tasting room

Viña Tarapacá was originally named Viña de Rojas in honor of its owner, Francisco de Rojas y Salamanca, and its vineyards were planted to French varieties. In 1892, Manuel Zavala-Meléndez acquired the winery and changed the name to Viña Tarapacá after the Chilean province of the same name. The Compañía Chilena de Fósforos, an important Chilean forestry enterprise and manufacturer of matches, acquired the winery and then invested in a 6,425-acre Maipo estate with 1,500 acres of vines to expand its fine-wine offerings. The winery was sold to VSPT in 2008.

Viña Tipaume

Founded: 1996

Address: Cerrillos, Cachapoal, http://tipaume.cl

Owner: Pouzet-Grez family

Winemaker: Yves Pouzet

Viticulturist: Yves Pouzet

Known for: Carmenère

Signature wine: Tipaume

Other labels: Grez

Visiting: Open to the public by appointment only; tasting room

Yves Pouzet, a French viticulturist and winemaker, has been living in Chile for decades and producing very small lots (fewer than two hundred cases annually) at his sixteen-acre estate, Tipaume. He sells most of his grapes to Cono Sur to supply the cash needed to run his micro-winery. His "Tipaume," from the Alto Cachapoal, is a blend of 60 percent Cabernet Sauvignon, 30 percent Carmenère, 5 percent Merlot, 3 percent Lacrima Cristi, and 2 percent Viognier, made of grapes from his certified organic vineyard. Tipaume strictly follows biodynamic practices and is a MOVI member.

Viña von Siebenthal

Founded: 1998

Address: O'Higgins sin número, Panquehue, Aconcagua, www.vinavonsiebenthal.com

Owner: Mauro von Siebenthal

Winemaker: Mauro von Siebenthal

Consulting winemaker: Stefano Gandolini

Viticulturist: Mauro von Siebenthal

Known for: Carmenère, Petit Verdot, Cabernet Sauvignon, Syrah

Signature wine: Parcela 7

Other labels: Carabantes, Montelìg, Carmenère, Toknar, Tatay de Cristobal

Visiting: Open to the public by appointment only; tasting room

Swiss-born Mauro von Siebenthal is a former lawyer who, after working for twenty-five years in Switzerland, started to look for vineyard land locally to indulge a long-held passion for wine. However, his desire to plant new varieties was not easy to satisfy in Switzerland. After tasting some wines from Errázuriz and receiving some pictures of Aconcagua from a friend, he was off to Chile. Mauro is dedicated to making top-quality wines from three vineyards in Panquehue at his winery, which, coincidentally, sits adjacent to Errázuriz. A MOVI member winery, von Siebenthal makes six distinctive wines, all red.

Viñedo Chadwick

Founded: 1992

Address: Av. Nueva Tajamar 481, Torre Sur, Oficina 503, Las Condes, Santiago, www.vinedochadwick.cl

Owner: Eduardo Chadwick

Winemaker: Francisco Baettig

Viticulturist: Ricardo Rodríguez

Known for: Cabernet Sauvignon, Cabernet Franc, Carmenère

Signature wine: Viñedo Chadwick

Visiting: Open to the public by appointment only

Viñedo Chadwick, in Puente Alto (Alto Maipo), was originally used as a family home. Eduardo's father, Alfonso Chadwick Errázuriz, was an ardent polo player, like so many of Chile's elite, and a piece of the sixty-two-acre property was used for this purpose. Eduardo came to live here in 1992, the year before his father died, and decided to turn the polo field into a vineyard planted to Cabernet Sauvignon, Merlot, and Cabernet Franc. The winery produces a single wine, of which the 1999 bottling was the initial release. It is considered one of Chile's great red wines and is a perennial high scorer in well-regarded wine publications.

Viñedos Marchigüe (Errázuriz Ovalle)

Founded: 2002

Address: Amunátegui 178, Piso 5, Santiago, www.eov.cl

Owner: Francisco Javier Errázuriz Talavera

Winemaker: Javier Regules

Viticulturist: Victor Cares

Known for: Cabernet Sauvignon

Signature wine: Panilonco

Other labels: Panul, Tierruca, Veo, Mahuida, Canto de Flora, Tricyclo

Visiting: Open to the public by appointment only; tasting room

After the Errázuriz family left the wine business many years ago, Eduardo Chadwick bought the original Errázuriz vineyard and brand to establish the well-known Errázuriz wine company. In 1992, when Francisco Errázuriz decided to get back into wine, a new company, with the confusing name of Errázuriz Ovalle, was born. Francisco is a well-known figure in Chilean business and politics who ran on the right-wing ticket in the presidential elections of 1989, when Augusto Pinochet was stepping down. Errázuriz runs a multinational company with interests in mining, agriculture, car distribution,

forestry, and supermarkets. The 7,400-acre Marchigüe wine estate is one of his most recent activities. For its first seven years, it sold grapes to Montes, Santa Carolina, and Veramonte. Since then it has become a wine producer that claims to be the second largest private producer in Chile, twice the size of Viña San Pedro.

Vinos del Sur (VinSur)

Founded: 1994

Address: Fundo Las Cañas, Camino El Morro sin número, San Javier, Maule, www.vinosdelsur.cl

Owners: José Esturillo, Francisco Gillmore

Winemaker: Patricia Inostroza

Consulting winemaker: Andrés Sánchez

Known for: Chardonnay, Cabernet Sauvignon, Merlot, Syrah, Cabernet Franc

Signature wines: Carpe Diem Tierra Roja, Rucahue Gran Reserve

Other labels: Concepción

Visiting: Open to the public by appointment only

The winery consortium is a partnership between Francisco Gillmore (of the namesake winery) and José Esturillo, whose family purchased a thousand acres of rugged, mountainous terrain in 1982. The consortium consists of Rucahue, the first Maule project; Carpe Diem, a Maule joint venture with the Fundación Chile, a government program that has also supported efforts by Viu Manent, Bisquertt, and others; and Concepción, a third Maule-based project with other investors. In 1999, Vinos del Sur acquired a new property in Itata's Larqui to produce everyday and premium reserve wines, working with Syrah, Carmenère, and Cabernet Franc.

Viu Manent

Founded: 1935

Address: Carretera del Vino, km 37, Colchagua, www.viumanent.com

Owner: Viu-Bottini family

Winemaker: Patricio Celedon

Consulting winemaker: Aurelio Montes

Viticulturist: Miguel Mujica

Known for: Malbec, Carmenère, Cabernet Sauvignon, Syrah

Signature wine: Viu 1

Other labels: El Incidente, ViBo (in Argentina), Secreto

Visiting: Open to the public; tasting room; restaurant (Rayuela Wine & Grill Restaurant)

In the mid-1930s, the Catalonian Viu-Manent family founded Bodegas Viu in Santiago, where they bottled and sold wine under the Vinos Viu brand. Some thirty years on, they acquired an estate of 371 acres of vines, a winery, and a manor house. Their holdings have expanded over time to include 627 acres in San Carlos, El Olivar, and La Capilla, the latter an area hard hit by the 2010 earthquake. Originally selling in bulk, the winery now produces premium wines and is lauded for its restaurant, the home base of celebrity chef Pilar Rodríguez, and its food and wine pairings. Viu Manent is one of the few wineries to be certified carbon neutral.

VSPT Wine Group

See Altaïr, Casa Rivas, Leyda, Misiones de Rengo, San Pedro, Santa Helena, Viña Mar, Viña Tarapacá

The VSPT Wine Group (Viña San Pedro Tarapacá Wine Group, www.vspt.cl) is the second-largest Chilean wine exporter and among the most important winemaking groups in the domestic market. In addition to its holdings in Chile, it owns Finca La Celia and Bodega Tamarí in Argentina. All are renowned cellars: each has its own distinctive brands, which are representative of the best wines these terroirs can deliver.

William Cole Vineyards

Founded: 1999

Address: Camino Tapihue, km 4.5, Casablanca, Aconcagua, www.wcv.cl

Owner: William Stevens Cole

Winemaker: Viviana Fonseca

Consulting winemaker: Peter Mackay

Viticulturist: Diego Campero

Known for: Chardonnay, Pinot Noir, Sauvignon Blanc, Cabernet Sauvignon

Signature wine: Bill Limited Edition

Other labels: Albamar

Visiting: Open to the public; tasting room

Wyoming born and a former U.S. Marine, William Cole left a very successful software career to pursue a dream. Cole's team consists of Viviana Fonseca and the antipodean consultant Peter Mackay, whose Southern Cross Wine Services advises wineries in Chile, China, and Argentina. The property consists of 319 contiguous acres. It has nothing to do with the winery of the same name in Napa Valley.

William Fèvre Chile

Founded: 1992

Address: Huelén 56, Oficina B, Providencia, Santiago, www.williamfevre.cl

Owner: Pino family

Winemaker: Felipe Uribe

Consulting winemaker: Alberto Antonini

Viticulturist: Nicolás Ianuzzi

Known for: Cabernet Sauvignon, Cabernet Franc, Pinot Noir, Carmenère, Merlot, Chardonnay, Sauvignon Blanc

Signature wine: Espino

Other labels: Antis, Chacai

Visiting: Open to the public by appointment only; tasting room

This winery was one of the pioneering French-Chilean joint ventures. William Fèvre, a noted Chablis producer, felt handcuffed by the strict French wine laws and wanted to expand. He settled in the Alto Maipo with a Chilean partner, Victor Pino, making wines from three riverbank terraces. Fèvre is now retired, and his company has been sold to the Champenois Henriot group. Most of his shares in the winery were purchased by the Pino family, who, not being wine specialists, brought in the consultants Pedro Parra and Alberto Antonini and a full-time winemaker, Felipe Uribe. The winery has a new project in Malleco that we will be hearing about down the road, especially for Pinot Noir.

5

BRAZIL

A gaucho next to the vineyard of Fortaleza do Seival in Candiota, Campanha. Photo by Silvia Tolon for IBRAVIN, Wines of Brazil.

Planted acreage: 227,300

World rank in acreage: 19

Number of wineries: 1,162 (approximately 160 focusing on fine wine)

Per capita consumption (liters)/ world ranking: 1.8/101

Leading white grapes: Chardonnay, Muscat, Niagara, Riesling Italico

Leading red grapes: Bordo, Cabernet Sauvignon, Concord, Isabella, Merlot, Pinot Noir (mostly for sparkling wine)

Memorable recent vintages: 2002, 2004, 2005, 2008, 2011, 2012

INTRODUCTION

Brazil conjures up a wealth of associations—soccer, beaches, Carnaval, samba, and the beautiful girl from Ipanema. Alas, wine is not on that list. Not much wine from Brazil is exported, and Brazilians themselves are far more likely to drink beer or *cachaça*.

Even so, after Argentina, Australia, South Africa, and Chile, Brazil is the fifth-largest Southern Hemisphere wine producer, vinifying some 84.5 million gallons annually. The grape and wine industries are important segments of Brazil's agricultural economy, employing tens of thousands of families. A typical household involved in viti- or viniculture works about five acres.

Brazilian wine is divided into two categories. The first is common (table) wine, made from American grapes (*Vitis labrusca*), mainly Isabella, Bordo, and Niagara. This accounts for three-quarters of all plantings and two-thirds of all production. The other is fine wine, based on *Vitis vinifera*, most of which is produced in the southernmost state of Rio Grande do Sul and concentrated in the region of Serra Gaúcha, home to about 60 percent of the nation's wine production. That said, grapes are farmed in seven different states across six climate zones. Brazilian wine country extends over almost 2,500 miles (roughly the distance from San Francisco to New York), from the equatorial vineyards in Pernambuco to the Uruguayan border of Campanha Gaúcha. There are three large wine companies (Miolo, Salton, and Aurora), four midsize houses (Chandon Brazil, Garibaldi, Perini, and Casa Valduga), and many small producers—close to eight hundred wineries in Serra Gaúcha alone.

In their choices of fine wine, Brazilians are decidedly unpatriotic, with domestic wines making up less than 20 percent of sales. Ironically, less than twenty years ago the reverse was true. From 2003 to 2013, the value of wine imported into Brazil tripled.* Why would that be? First, domestic wines are taxed at twice the rate of imports (55 percent and 27 percent, respectively, with tax rates even lower for wine from Mercosur countries). Second, draconian drinking-and-driving laws are a damper on all alcoholic consumption. As in the United Kingdom, 90 percent of the country's wine is sold at retail. In Brazil sales are driven by the Pão de Açuar grocery chain, the Tesco of Brazil, with six hundred stores. In an attempt to boost sales of domestic wine, the Brazilian wine industry reached an agreement with the grocery stores in October 2012: the stores committed to increasing shelf space for domestic wines from 15 percent to 25 percent, with the goal of doubling the low per capita consumption. At less than two liters per year per person, Brazil consumes significantly less wine than some Muslim countries, including the Maldives and the United Arab Emirates.

LOCAL HISTORY

Though grapes were grown and wine was made before then, the mid-1870s marks the real birth of the Brazilian wine industry. This was a period when scores of Italian immigrants settled in Rio Grande do Sul (notably in Serra Gaúcha). Seeking to attract settlers into the immense southern portion of this vast country, Brazil offered

* Christian Burgos, "The Brazilian Thirst for Wine," *Meininger's Wine Business International*, August 30, 2012, www.wine-business-international.com/156-bWVtb2lyX2lkPTQ1NSZtZW51ZV9jYXRfaWQ9--en-magazine-magazine_detail.html, accessed January 17, 2014.

land grants and welcomed boatloads of Italians from Trentino (mainly Vicenza) and the Veneto (mainly Breganze) seeking to escape the poverty of their homeland and begin new lives. Other immigrants arrived from Germany, Portugal, and Poland. Simultaneously, of course, boats were headed to North America, notably New York's Ellis Island, and immigrants boarding ships were often unaware of their final destination. It is claimed that the Gallo and Miolo families, two of the Americas' leading wine clans, were originally from the same surrounding area of northern Italy, and it's purely by chance that one family ended up in Brazil and the other in the United States.

While the Germans mostly settled around Porto Alegre and south of Serra Gaúcha, the Italians made their way to the area around Bento Gonçalves, the heart of wine country, likely drawn by its resemblance to their homeland and the promise of rich agriculture. Vines were brought over, along with other crops, and wines were made for home consumption. It became evident that European vines didn't perform consistently well in cooler and wetter southern Brazil. When North American native and hybrid grapes made their way to Brazil's wine heartland by way of global commerce, they performed better in the local climate and quickly became the primary cultivars. These varieties still represent the lion's share of Brazil's grape plantings, forming the backbone of the trade in wine, juice, concentrate, and table grapes.

Brazil's wine culture was unremarkable until winemaking families decided to get together and form cooperatives. The Garibaldi and Aurora co-ops, both started in 1931, were established to combat the economic hardships of the times while laying the foundation for more organized business. Today there are about twenty active co-ops in Brazil, five of them in Serra Gaúcha. Of these, the most prominent are still Aurora, with 1,100 member families, and Garibaldi, with 350.

By the mid-1960s, the multinational companies had arrived, seeking a foothold in the increasingly exciting Brazilian wine industry. Steered by the Italians (Martini and Rossi, and Cinzano) and the French (Moët & Chandon and Pernod), Brazilian producers were advised that to be taken seriously by discerning wine drinkers, they needed to improve quality. This meant producing *vinifera*-based bubblies (Chardonnay and Pinot Noir) and table wines focusing on Bordelais grapes. It was soon determined that the cooler, wetter climate of Serra Gaúcha favored Merlot, though growers also experimented with Cabernet Sauvignon and Cabernet Franc. The consensus is that the Carraro family was the first to plant Merlot and the Miolos the first to plant Cabernet Sauvignon.

Over time, Brazilian winemakers became better at sourcing and selecting grapes and matching them to terroir. Between the 1970s and the 1990s, a local fine-wine culture emerged. The industry expanded both south toward the border with Uruguay and into the semiarid northeast, in the São Francisco River Valley across the states of Bahia and Pernambuco. Back in Serra Gaúcha, led by wineries such as Dal Pizzol, Casa Valduga, Salton, and Miolo, quality production was gradually expanding in small and moderate-scale operations. During the 1980s and early 1990s, as political and economic circumstances challenged growers, the time-honored practice of selling almost exclusively to the co-ops began to erode. Some growers began to sell to the emerging wineries, while others decided to make a go of it themselves. By the late 1980s, many small wineries and grape growers were investing heavily in developing quality operations. Some sent their children abroad to take enology courses, work harvests, and learn about global palates and practices. These shifts gave rise to the contemporary era of Brazilian wines. Today, many quality-focused, artisan and boutique family *adegas* (cellars and wineries) have joined the successful, now-larger family operations and modern co-ops, and together they contribute to a vibrant wine culture. As a capstone to this effort, the Brazilian Wine Institute (IBRAVIN) was formed in 1998.

Since 2000, forward-thinking Brazilian producers have focused on selecting varieties suited to their climates and exploring regions previously either misunderstood (such as Santa Catarina and Paraná) or underappreciated (Serra do Sudeste and Campanha Gaúcha). Unlike Argentina and Chile, Brazil has achieved this success with minimal intervention from international consultants. Miolo worked with Michel Rolland but ended the relationship in 2012 (because, according to Anthony Darricarrère of Routhier & Darricarrère, "northern Italians don't like anyone telling them what to do").

BRAZILIAN GAME CHANGERS

ANTONIO DAL PIZZOL

An affable old-guard producer based in Faria Lemos, just outside Bento Gonçalves, Antonio is widely considered to be the patriarch of the boutique wine movement. While others in the late 1970s chose to expand, Antonio and his brother Rinaldo remained small, with a focus on direct sales to customers, both consumer and trade, and an emphasis on small lots and fewer mainstream varieties. Their offerings included varietal Trebbiano, Touriga Nacional, and Tannat. They were also the first in Brazil to bottle a varietal Cabernet Franc, in 1978. Antonio has almost a cult following in São Paulo, a testament to his emphasis on quality and his boutique vision: he sells over half of his wine without distributors. The Dal Pizzol winery, founded in 1974, has inspired several great boutique winemakers, including Luiz Milani (Milantino) and Ademir Brandelli (Don Laurindo).

MARIO GEISSE

The Chilean Mario Geisse first came to Brazil in 1976 with Moët & Chandon, charged with crafting quality sparkling wine in Brazil. This effort led him to the discovery of the Pinto Bandeira area of Serra Gaúcha, one of Brazil's first two officially recognized regions. Ultimately convinced that Moët & Chandon didn't share his goals for quality, Mario left and began independently producing sparkling wine in 1979, effectively demonstrating that world-class bubbly was possible in Brazil. Making limited amounts of *méthode traditionnelle* wines with different ranges of *en tirage* aging, Mario is credited with creating the quality sparkling-wine industry in Brazil. The lone Brazilian wine shown on Jancis Robinson's panel at China's Wine Future show in 2011 was Mario's Terroir Brut 1998 in magnum. Retaining his connections to Chile, Mario is also senior winemaker and technical and enological director of the highly regarded Casa Silva winery in Colchagua.

ADRIANO MIOLO

Many family wineries are clustered in the Vale dos Vinhedos, but Miolo, led by the tireless Adriano, is different. With the same visionary outlook and focus on quality shown by neighboring Argentina's Nicolás Catena, Adriano went through technical courses in enology (including time in Mendoza), and obtained a master's degree in marketing wine from the Bordeaux International Wine Institute, a four-year endeavor involving study in twelve countries. Today, under Adriano's leadership, Miolo is the largest grower and producer of fine wine in Brazil. It is credited with producing iconic wines: the celebrated "Lot 43," named for the original land grant given to the Miolo family, is a Bordeaux-style red blend, heavy on the Merlot. The winery is also known for pioneering new areas (Miolo was among the first to venture into Campanha Gaúcha) and providing employ-ment to hundreds of families across the country (as they operate wineries in multiple regions), while demonstrating that scale and quality are not mutually exclusive.

FLAVIO PIZZATO

A leading protagonist of the "new wine movement" of the late 1990s and early 2000s, Flavio is an electrical engineer who returned to the family winery after the family left a co-op. His father, Plinio, one of the largest growers of the French varieties brought to Brazil by international ventures during the 1970s, was angry that his high-quality grapes were being lost in blends. In 1997, Flavio and his brother Ivo aggressively moved the family onto a new path with a focus on improving viticulture, and in 1999, their Pizzato Merlot became a benchmark for the DO red wines of Vale dos Vinhedos. Back in 1985, the Pizzatos were the first growers in Serra Gaúcha to move away from training vines on traditional pergolas and adopt vertical shoot positioning and the first winemak-ers to vinify less common varieties like Egiodola, Tannat, and Alicante Bouschet for quality and not purely for blending.

REGIONS AND WINERIES

Brazil's primary wine regions encompass a huge range of climates, altitudes, tempera-tures, precipitation zones, and geologies. Wine is made in Campanha Gaúcha, in the far south, at altitudes around 700 feet, and in the São Francisco Valley, the northernmost region, at close to 1,200 feet. Brazil's wine industry has shown the ability to make wines of Old World structure with New World flavor: it grows a cornucopia of diverse and uncommon grapes, produces plentiful and good sparkling wine, and has an abundance of newly popular Muscat, known in Brazil as Moscatel. The downsides include inconsis-tent vintages, high costs and taxes, and, for exporters, a complicated language and lack of international recognition. I can't do much about the first three, but I can certainly help address the last point!

VALE DO SÃO FRANCISCO

Sitting between 8° and 9° south, the São Francisco Valley spans the two northeastern states of Bahia (near the southern city of Casanova) and Pernambuco (near the northern

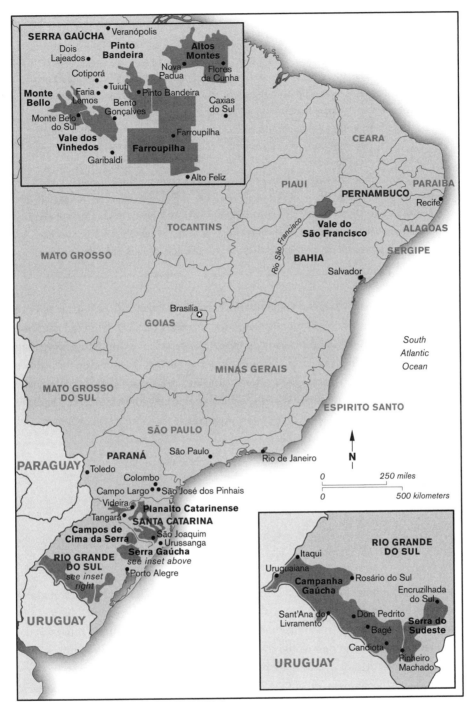

Inset (top left):

SERRA GAÚCHA

Veranópolis

Dois
Lajeados

Pinto
Bandeira

Altos
Montes

Cotiporã

Nova
Padua

Flores
da Cunha

Monte
Bello

Faria
Lemos

Tuiuti

Pinto Bandeira

Caxias
do Sul

Bento
Gonçalves

Monte Belo
do Sul

Vale dos
Vinhedos

Farroupilha

Farroupilha

Garibaldi

Alto Feliz

Main map:

CEARA

PIAUI

PERNAMBUCO

PARAIBA

Recife

Rio São Francisco

Vale do
São Francisco

ALAGOAS

SERGIPE

TOCANTINS

BAHIA

MATO GROSSO

Salvador

Brasília

GOIAS

South
Atlantic
Ocean

MINAS GERAIS

MATO GROSSO
DO SUL

ESPIRITO SANTO

SÃO PAULO

PARANÁ

São Paulo

N

Rio de Janeiro

PARAGUAY

Toledo

0   250 miles

Colombo

0   500 kilometers

Campo Largo

São José dos Pinhais

Videira

Planalto Catarinense

Tangará

SANTA CATARINA

Campos de
Cima da Serra

São Joaquim

Urussanga

Serra Gaúcha
see inset above

RIO GRANDE
DO SUL
see inset
right

Porto Alegre

URUGUAY

Inset (bottom right):

RIO GRANDE
DO SUL

Itaqui

Uruguaiana

Rosário do Sul

Campanha
Gaúcha

Encruzilhada
do Sul

Sant'Ana do
Livramento

Dom Pedrito

Serra do
Sudeste

Bagé

Candiota

URUGUAY

Pinheiro
Machado

Map 8. Brazil's wine regions

municipalities of Lagoa Grande and Santa Maria da Boa Vista). Of the 31,000 acres of grapes planted, 29,600 are in *labrusca* varieties, intended for table fruit and common wine. The high humidity, a rainless and perennially warm climate, and ease of irrigation mean that winegrowing is not difficult. Some wineries harvest twice or even three times per year. Since the mid-2000s, new efforts with European *vinifera* grapes, especially Syrah, have shown that quality wines can be made in ripe and quaffable styles. There are fewer than ten wineries in the region, with only a couple focusing on fine wine.

RECOMMENDED PRODUCERS

Ouro Verde (*see* Miolo Wine Group)
Vinibrasil (Vitivinícola Santa Maria)

PARANÁ

In south-central Brazil, Paraná is the only state to have shown the potential for making fine wine, although common wine is also produced in the states of Minas Gerais, São Paulo, and Mato Grosso. In Paraná, home of the immense and stunning Iguazu Falls, winemaking is concentrated in the districts of Colombo, São José dos Pinhais, and Campo Largo. Its origins date back to the Italian colonization of the late 1800s. It is considered the northernmost state in Brazil's southern winemaking zone and the youngest with respect to producing fine wine. Various diseases devastated its vineyards in the 1960s. Viticulture was reintroduced into northern Paraná in the 1970s by the Brazilian Japanese community, who focused on producing fine-wine grapes. To compensate for both a shortage of grapes and a shortage of labor, however, grapes are frequently brought in from nearby Rio Grande do Sul. In the west, near Toledo, vines thrive in rich soils at altitudes up to six hundred feet. This higher altitude results in moderate thermal amplitude, though median annual temperatures hover around 66°F. Cabernet Sauvignon and Chardonnay seem to be the region's most exciting fine wines so far.

RECOMMENDED PRODUCER

Dezem

SANTA CATARINA

Situated north of Rio Grande do Sul, the state of Santa Catarina is home to higher-altitude and cooler-climate winemaking. With elevations ranging between 3,000 and 4,600 feet and average temperatures of 49°F in the winter and 71°F in the summer, the coolest spot in Brazil is Santa Catarina's city of Urubici, 35 miles outside São Joaquim, where most of the *vinifera* wineries are situated. The climate is like the little girl with the curl: when it is good it is very, very good, and when it is bad it is horrid. Santa Catarina can be best described as winemaking on the edge. The cold comes early, and frost is an enormous problem: only one in five vintages is excellent. As befits the occasionally frigid territory, São Joaquim is home to South America's only real ice wines. High-volume wine production is centered on the Vale do Rio do Peixe, around the cities of Videira, Tangará, and Pinheiro Preto. More grapes are grown in the southern district near Urussanga, in Pedras Grandes and Morro da Fumaça, and in the Vale do Rio Tijucas, chiefly in the municipalities of Nova Trento and Major Gercino. In both these regions, viticulture is based on *labrusca* and hybrid varieties. The total vineyard area in Santa Catarina is ten thousand acres, with 20 percent planted in *vinifera*.

The very new wine subregion of São Joaquim is situated near the namesake town and the adjacent municipalities of Campos Novos and Caçador. Santa Catarina's smallest subregion, it is uniquely planted 100 percent to *vinifera*. Sited on an elevated plateau, it has been called the Planalto Catarinense. Although it is still establishing its reputation, São Joaquim seems to excel in whites, notably Chardonnay, and lovely sparkling wine. In years when the reds do ripen, they have excellent structure that bodes well for aging potential.

RECOMMENDED PRODUCERS

Kranz

Panceri

Pericó Vinhos

Quinta Santa Maria

Sanjo

Suzin

Villa Francioni

Villaggio Grando Boutique Winery

Vinícola Santa Augusta

Campos de Cima da Serra and Serra Gaúcha

Rio Grande do Sul, Brazil's southernmost state, is where 90 percent of all Brazilian wine is made. Campos de Cima da Serra is a cool-climate area that has been known more for its superb apples than for its wine. With vineyards situated at 3,200 feet, this newer region's wines are renowned for their acidity and good color. Conditions here are quite similar to those of Santa Catarina's São Joaquim. The main town is Vacaria, home to Brazil's largest and best-known agro-nursery, Rasip Agro Pastoril SA, which produces and sells apples, grapes, cheese, and other dairy products.

Undeniably Brazil's most important wine region, the Serra Gaúcha sits some sixty miles south of Vacaria and seventy-five miles north of Porto Alegre, the state capital and its largest city. With close to 76,000 acres under vine, it has approximately eight hundred wine companies (with a moderate number focusing exclusively on *vinifera*). It receives over two hundred thousand visitors annually, making it Brazil's equivalent of the Napa Valley. Incorporating the Vale dos Vinhedos, Brazil's lone *denominação de origem* (DO), and Pinto Bandeira, an *indicação de procedência* (IP) renowned for sparkling wine, the Serra Gaúcha is the epicenter for quality wine. Located at 29° south, approximately level with Argentina's San Juan, Serra Gaúcha's wine zone sits between elevations of 1,900 and 2,600 feet, with annual rainfall of about 67 inches and a median temperature of 63°F. It looks and feels like much of Western Europe's French and Italian wine country, although it is far more humid. As in Europe, it can rain at harvest time. The cooler climate means that wines rarely exceed 13.5 percent alcohol. Reds are usually right around 13 percent and whites about 12.5 percent. These alcohol levels produce a European style but with a ripe fruit profile—a perfect cross of Old and New worlds.

Eighty percent of the region's production is in American native and hybrid varieties, mainly Isabella, Bordo, Niagara, and Concord. Among *vinifera* varieties, whites are led by Muscat, Chardonnay, and Riesling Italico. The main red varieties are Merlot, Cabernet Sauvignon, Cabernet Franc, and Tannat. Pinot Noir is also grown, mostly for sparkling wine. Merlot is the longest-established *vinifera* grape and produces the most consistent results.

The heart of the region is the Vale dos Vinhedos. Situated in and around the municipality of Bento Gonçalves, vineyards here are the most expensive in Brazil, though still a good value when compared to the Napa Valley or Burgundy. Encompassing a territory of fifty square miles, a quarter of which is planted, the soils are basaltic, similar to those of southeastern Australia and America's Pacific Northwest. As an official DO, Vale dos Vinhedos has a variety of labeling restrictions. Only Merlot and Chardonnay may be bottled with the specific appellation, and they may be blended only with certain sanctioned grapes. There are also minimum alcohol levels, restrictions on the grape varieties and techniques used for sparkling wine (only *méthode traditionnelle* production is allowed), and stipulations on bottle and barrel aging (no oak chips) and production techniques. All grapes used in Vale dos Vinhedos–labeled wines must come from the appellation. The Vale dos Vinhedos specializes in Bordeaux red grapes, especially Merlot and Cabernet Franc, in addition to Chardonnay and sparkling wines of myriad styles and varieties. Though Cabernet Sauvignon, Tannat, Touriga Nacional, and other warmer-climate grapes are cultivated, they excel only in riper vintages. Geographically, the Vale dos Vinhedos encompasses Bento Gonçalves, part of Monte Belo do Sul, and the municipality of Garibaldi, known for its very good Chardonnay. With its well-developed wine-tourism attractions (including a wine-tourism train that runs three times daily), it is worth a visit.

Just north of Vale dos Vinhedos is the district of Faria Lemos, encompassing the smaller Vale Aurora, which is essentially a continuation of the Vale dos Vinhedos. Faria Lemos excels at producing a range of wines. The Vale Trentino is smaller and is situated near the town of Farroupilha, at 2,600 feet, the source of the best and close to 50 percent of the country's Muscat. Pinto Bandeira sits at a slightly higher altitude and is recognized for its sparkling wine and Chardonnay. Other significant wine areas include those north of the Antas River (Cotiporá, Veranópolis, Tuiuti, and Dois Lajeados) as well as Caxias do Sul and Altos Montes, denoted as an IP in early 2013, with its key subareas of Flores da Cunha and Nova Pádua. Finally, heading south toward Porto Alegre, you arrive in Alto Feliz.

RECOMMENDED PRODUCERS

Aurora

Basso Vinhos e Espumantes

Boscato

Casa Valduga

Cavalleri

Chandon (*see* Chandon Argentina)

Courmayeur

Dal Pizzol

Dom Cândido

Don Giovanni

Don Laurindo

Garibaldi

Geisse

Grupo Decima

Laurentia

Lidio Carraro

Miolo Winery (*see also* Miolo Wine Group)

Perini

Peterlongo

Pizzato

Salton

Serra do Sudeste

About three hours south of Bento Gonçalves by car, the wine region of Serra do Sudeste is equidistant between the northern Serra Gaúcha and the southern Campanha Gaúcha. Although this area had experimental plantings in the late 1990s, Lidio Carraro is credited with pioneering the region in 2000 (instead of heading to Campanha Gaú-cha, as most other winemakers were doing). Two areas form the heart of wine country: Encruzilhada do Sul and the less significant Pinheiro Machado to its southwest. Characterized by rolling hills and flat plains at about 2,500 feet above sea level, Encruzilhada do Sul has a moderate climate, with dry summers and slightly warmer temperatures than the Serra Gaúcha. What makes the region special is the soil, a combination of limestone and granite that is low in organic nutrients. Relatively small, at just six thousand acres, the area is best known for Chardonnay, Pinot Noir, and Tannat. However, there is excitement about Portuguese red grapes (Touriga Nacional and Alicante Bouschet) and lesser-known Italian red varieties, such as Ancellotta. There are fewer than ten producers growing and sourcing fruit here, and no wineries: the fruit is all trucked north to Serra Gaúcha.

Campanha Gaúcha

The most significant buzz in Brazilian wine today is centered on Campanha Gaúcha, on the border with Uruguay (often referred to as Fronteira [the border] or simply Campanha [the countryside]). Although the region was historically known for growing rice and raising cattle, Campanha's warmer and drier climate, high thermal amplitude, and clay and granite soils were identified in the early 1970s as the elements of a winemaking paradise. The terrain also offers the potential for mechanization in the vineyards (necessary because of labor shortages) and ample land for expansion.

Campanha Gaúcha, covering an area the size of Belgium, is made up of three distinctive subregions that become cooler as you move east and slightly north: Uruguaiana and Itaqui; Sant'Ana do Livramento (including Palomas) and Rosário do Sul; and Dom Pedrito, Candiota, and Bagé. Fewer than twenty producers are working 3,500 acres of 100 percent *vinifera* grapes, most prominently 900 acres of Cabernet Sauvignon. In 2012, 15 percent of Brazil's fine wine emanated from Campanha Gaúcha, and that percentage is expected to grow.

In Campanha, there are records of secular vineyards (as opposed to those established much earlier by missionaries) dating back to 1881 and wines dating back to 1886, but winemaking was dormant in the area until 1973, when Harold Olmos of UC Davis identified Campanha as the most viable, expandable location for quality winemaking in Brazil and traveled there to assist in National Distillers' Almadén project. Four years later the first vineyards, with twenty-two different varieties (eleven white and eleven red), were planted in Sant'Ana do Livramento. Subsequently, the Miolo family established the Fortaleza do Seival vineyard, close to Candiota, and the Valduga and Salton families followed suit. Today, however, just two wineries (Almadén and Seival) share 70 percent of the region's total production.

Though it is too early to draw definitive conclusions, the consensus is that the higher clay content of the soil in eastern Bagé results in more aromatic grapes. Although Cabernet Sauvignon and other red grapes—including Syrah, Tannat, Touriga Nacional, and Tempranillo—dominate in this warmer and less humid environment, several people I spoke with believe that the whites, especially Viognier, are underrated. Campanha is sunnier than some of the more northerly wine areas, mitigating slightly the diurnal temperature extremes resulting from its southerly latitude.

RECOMMENDED PRODUCERS

Campos de Cima

Cordilheira de Sant'Ana

Dunamis Vinhos e Vinhedos

Routhier & Darricarrère

Seival (*see* Miolo Wine Group)

WINERY PROFILES

One of Casa Valduga's vineyards in the Serra Gaúcha of Brazil. Photo from the archives of Casa Valduga winery.

Aurora

Founded: 1931

Address: Olavo Bilac 500, Bento Gonçalves, Rio Grande do Sul, www.vinicolaaurora.com.br

Owners: 1,100 member families

Winemaker: Nauro J. Morbini

Consulting winemaker: André Peres Jr.

Viticulturists: 1,100 viticulturist member families, with João Carlos Rigo heading the agricultural technical department

Known for: Merlot, Chardonnay, Muscat

Signature wines: Aurora Moscatel, Pequeñas Partilhas Cabernet Franc

Other labels: Marcus James, Conde de Foucauld, Clos des Nobles, Saint Germain, Maison de Ville, Casa de Bento

Visiting: Open to the public by appointment only; tasting room

Even if you think you have never tasted a Brazilian wine, you may well have sampled Aurora's offerings. Marcus James, now a Constellation wine brand imported from Argentina, began its life in the 1980s as the Aurora co-op's brand. The very un-Brazilian name was derived from the first names of an Aurora executive and the son of his American business partner. Aurora still sells Brazilian-made Marcus James wine domestically. A co-op of 1,100 families dating back almost eighty years, Cooperative Vinícola Aurora was receiving grapes from more than 1,500 family growers at its peak in the mid-1990s. While it has since cut back on volume, it still makes close to thirty different lines of wine that range from basic to well-made small-lot offerings. To get a sense of the scale of the operation, visit the winery's website and browse through their selections. The winery, situated smack in the center of Bento Gonçalves, receives thousands of visitors annually by appointment.

Basso Vinhos e Espumantes

Founded: 1974

Address: Monte Bérico, Second District, Farroupilha, Rio Grande do Sul, www.vinicolabasso.com.br

Owners: Moacir Basso, Rudimar Basso, Rui Basso, Simão Basso

Winemaker: Magnos Basso

Consulting winemaker: Marcos Vian

Known for: Muscat, Pinot Noir

Signature wine: Monte Paschoal Dedicato Pinot Noir

Other labels: Del Grano

Visiting: Open to the public

In 1940, Hermindo Basso founded a family-owned winery, originally named Rural Cantina, in Monte Bérico, a subdistrict of Farroupilha. Today the family owns twenty-five acres and subcontracts with growers to work another 1,340 acres. Their premium offerings, under the "Monte Paschoal" label (named for a local mountain), are wide-ranging. Their specialty,

however, is Farroupilha's favorite grape, Muscat, which is offered up in five different wines including two sparkling wines, two *frizzantes,* and one dry table wine. In any given year they have at least sixteen different lots of Muscat fruit. That's commitment! The Pinot Noir is a promising passion of theirs.

Boscato

Founded: 1983

Address: VRS 314, km 12.5, Nova Pádua, Rio Grande do Sul, www.boscato.com.br

Owners: Clóvis Roberto Boscato, Valmor João Boscato

Winemaker: Clóvis Roberto Boscato

Viticulturist: Valmor João Boscato

Known for: Merlot, Cabernet Sauvignon, Ancellotta, Alicante Bouschet, Pinot Noir, Chardonnay, Refosco

Signature wine: Gran Reserva Merlot

Other labels: Anima Vitis

Visiting: Open to the public; tasting room

Located on the plateau of the Vale do Rio das Antas, Boscato, led by viticulturist Valmor Boscato, winemaker Clóvis Roberto Boscato, and his agronomist daughter Roberta, is one of the preeminent Brazilian *adegas.* With three wines in the top thirteen offerings of the 2012–13 wine guide *Guia adega vinhos do Brasil,* the Boscatos clearly know what they are doing. Specializing in red wine, the winery currently exports to Europe and Australia.

Campos de Cima

Founded: 2002

Address: Rua Bento Gonçalves 1225, Itaqui,

Rio Grande do Sul, www.camposdecima
.com.br

Owners: Hortência Ayub, José Ayub

Winemaker: Celito Guerra

Viticulturist: Antonio Santim

Known for: Tannat, Ruby Cabernet, Shiraz,
Chardonnay, Malbec

Signature wines: Tannat, Ruby Cabernet,
Espumante Brut

Visiting: Open to the public; tasting room

While the company has been around for over a decade, the Campos de Cima winery only opened at the end of 2012. Until then, all grapes were custom crushed elsewhere. The Ayub family has thirty-seven acres in Itaqui and is so passionate about winemaking that they have taken courses through the Wine and Spirits Education Trust and have the goal of becoming Masters of Wine. Few wineries specialize in Ruby Cabernet, a cross between Carignan and Cabernet Sauvignon, but this is Campos de Cima's defining variety, and this wine sells out first every year. (Note: the winery, located in Campanha, should not be confused with the northern area of Rio Grande do Sul called Campos de Cima da Serra.)

Casa Valduga

Founded: 1973

Address: Via Trento 2355, Vale dos Vinhedos, Bento Gonçalves, Rio Grande do Sul, www.casavalduga.com.br

Owners: João Valduga, Juarez Valduga,
Erielso Valduga

Winemaker: João Valduga

Viticulturist: João Valduga

Known for: Merlot, Cabernet Franc,
Chardonnay, sparkling wines

Signature wines: Storia, María Valduga
Sparkling Wine

Other labels: Casa Madeira, Domno, Villa
Valduga

Visiting: Open to the public; tasting room;
restaurants (María Valduga and Luiz
Valduga)

The Valduga family is one of Brazil's most venerable. Three generations work for the company, which traces its roots to Alto Adige. Four brothers came over together from Italy in 1875, with one each going to Buenos Aires, Montevideo, Santos, and Porto Alegre. The Valdugas sold fruit before starting their winery, which has become one of Brazil's leading estates. They have two successful restaurants, one more traditional and one modern, and the country's most professional visitor center and tasting room, which receives over eighty thousand visitors annually. They split their production evenly between sparkling wines and quality table wines. The newer Domno sparkling wines are made by the Charmat method and, together with the *méthode traditionnelle* Valduga bubbly, make the winery Brazil's fifth-largest sparkling-wine producer.

Cavalleri

Founded: 1987

Address: RS 444, km 24, Vale dos
Vinhedos, Bento Gonçalves, Rio Grande
do Sul, www.cavalleri.com.br

Owner: Nilso Cavalleri

Winemaker: Lorenço Vaccaro

Viticulturist: Nilso Cavalleri

Known for: Merlot

Signature wine: Merlot Reserva

Visiting: Open to the public

With its colonial-style building that evokes the atmosphere of Italy, Cavalleri is very welcoming to visitors. And its location on the main road (RS 444), between beautiful Bento Gonçalves and Monte Belo do Sul, makes it a convenient destination. Currently, the Cavalleri Winery has a production area of eighty acres and specializes in sparkling wines and Moscatel, though it produces a range of wines, including one of Brazil's few Carmenères.

Cordilheira de Sant'Ana

Founded: 2000

Address: Vila Palomas sem número, Sant'Ana do Livramento, Rio Grande do Sul, www.cordilheiradesantana.com.br

Owners: Gladistão Omizzolo, Rosana Wagner

Winemakers: Gladistão Omizzolo, Rosana Wagner

Known for: Chardonnay, Gewürztraminer, Sauvignon Blanc, Merlot, Cabernet Sauvignon, Tannat

Signature wine: Cordilheira de Sant'Ana

Other labels: Reserva dos Pampas

Visiting: Open to the public

Cordilheira de Sant'Ana is owned by a couple who are both winemakers. They met when they were working in the spirits industry and Gladistão Omizzolo was Rosana Wagner's boss. The winery name is based on its foothill (*cordilheira*) location not far from the frontier city of Sant'Ana

do Livramento; the border with Uruguay is about twenty minutes away by car. One of the only wineries practicing native-yeast fermentation exclusively, Cordilheira de Sant'Ana sells more than 60 percent of its wine direct at the winery. It is unusual in releasing its wines to market only when the winemakers consider them ready: when I was there in 2012, the current offering was the 2008 Merlot.

Courmayeur

Founded: 1976

Address: Av. Garibaldina 32, Garibaldi, Serra Gaúcha, Rio Grande do Sul, www.courmayeur.com.br

Owner: Mário Verzeletti

Winemaker: Gilson Berselli

Consulting winemaker: Firmino Splendor

Known for: Chardonnay, Pinot Noir

Signature wine: Retrato Courmayeur

Visiting: Open to the public by appointment only, tasting room

Named for an Italian commune in the Valle d'Aosta at the foot of Mont Blanc, on the border with France, this winery's offerings were among the first I tasted on an early trip to Brazil, and they left a distinct impression. Courmayeur was originally part of the Cinzano group but went private in 2003. Sparkling wine accounts for 90 percent of the winery's production. About half of its volume is the benchmark sparkling Moscatel, which, like all of Courmayeur's other wines, is made using the Charmat method.

Dal Pizzol (Vinícola Monte Lemos Ltda)

Founded: 1974

Address: RS 431, km 5, Faria Lemos,
Bento Gonçalves, Rio Grande do Sul,
www.dalpizzol.com.br

Owners: Antonio Dal Pizzol, Rinaldo Dal
Pizzol, Valdair Dal Pizzol, Valter Dal
Pizzol

Winemaker: Dirceu Scottá

Known for: Touriga Nacional,
Cabernet Sauvignon, Gamay,
Merlot, Ancellotta, Chardonnay,
Sauvignon Blanc, Tannat

Signature wines: Touriga Nacional,
Ancellotta

Other labels: Do Lugar

Visiting: Open to the public; tasting room

The Dal Pizzol family runs one of the
few Brazilian wineries that was established with a focus on *vinifera* grapes.
Whereas most wineries did not begin in
earnest with quality fruit until the 1980s
and 1990s, the Dal Pizzols understood
from the beginning that quality was the
future. Their first vintage, in 1978, commemorated the centenary of the family's arrival from Treviso. Given the small
size of the winery, its line of more than
a dozen wines may seem excessive, but
the family thrives on experimentation.
Well worth the visit, the winery has a museum with old winemaking equipment,
picnic grounds, a lakeside meeting space,
a play area, and an archival vineyard with
over 160 varieties planted. These grapes
are used to make one or two inimita-
ble blends each year, available only at
the winery.

Dezem

Founded: 2005

Address: Concórdia do Oeste, Toledo,
Paraná, www.dezemvinhosfinos.com.br

Owner: Susan Dezem

Winemakers: Anderson Schmitz, Marcos
Vian

Viticulturist: Marcos Link

Known for: Chardonnay, Cabernet
Sauvignon, Merlot, Cabernet Franc, Tan-
nat, Tempranillo, Pinot Noir, Sauvignon
Blanc, Malvasia

Signature wine: Atmo Chardonnay

Other labels: Extrus, Magne

Visiting: Open to the public by
appointment

An island of superiority, Dezem stands
out as the lone quality-focused winery in
Paraná. Since 1990, all of its fruit has been
estate grown, and since 2004, winemaking has been overseen by the very talented
Anderson Schmitz, who also works with
Sanjo and thirteen other wineries across
the country. Dezem is unique in Paraná in
focusing only on *vinifera* varieties, which
are used to make both sparkling and still
wines. In addition to the expected grapes,
Dezem has plantings of Negro Amaro,
Sangiovese, and Tempranillo.

Dom Cândido

Founded: 1875

Address: Linha Leopoldina sem número,
Vale dos Vinhedos, Bento Gonçalves, Rio
Grande do Sul, www.domcandido.com.br

Owner: Carlos Alberto Valduga

Winemaker: Daniel de Paris

Known for: Cabernet Sauvignon, Merlot, Marselan, Tannat, Malbec, Chardonnay

Signature wine: Gran Reserva Cabernet Sauvignon

Visiting: Open to the public; tasting room; restaurant

Carlos Albert Valduga grew up in the wine trade, assisting his father when he was a boy, and then set off to open his own operation, now a boutique winery. Favoring a smaller scale than the family winery, Casa Valduga, Dom Cândido produces a number of wines in small quantities. A pioneer in Cabernet Franc, Dom Cândido was also the first winery in Brazil to produce a varietal Marselan (in 2003), and the *4ª geração* (fourth-generation) bottling was named the best in the country in 2013 by *Anuário Vinhos do Brasil*. The modest restaurant is a nice complement to the winery but not as ambitious as that of Casa Valduga.

Don Giovanni

Founded: 1982

Address: Linha Amadeu, km 12, Pinto Bandeira, Bento Gonçalves, Rio Grande do Sul, www.dongiovanni.com.br

Owner: Ayrton Luiz Giovannini

Winemaker: Luciano Vian

Viticulturist: Luciano Vian

Known for: Chardonnay

Signature wine: Don Giovanni Brut sparkling wines

Visiting: Open to the public; tasting room; restaurant; inn

This family-owned winery was built on the former site of an early-twentieth-century brewery. A quaint, family-run bed

and breakfast with eight guestrooms and an award-winning restaurant have been added. The winery's focus is on sparkling wines (which account for 70 percent of the output), and the NV Stravaganza, which spends a year *en tirage,* is a great value. All their bubblies are hand riddled. Luciano Vian, a soft-spoken and talented winemaker, has been with the winery since its inception.

Don Guerino

Founded: 2004

Address: Rua dos Vinhedos sem número, Alto Feliz, Rio Grande do Sul, www.donguerino.com.br

Owner: Osvaldo Motter

Winemaker: Bruno Motter

Viticulturist: Osvaldo Motter

Known for: Ancellotta, Teroldego, Merlot, Tannat, Cabernet Sauvignon, Chardonnay, Moscato Giallo, Malbec

Signature wine: Gran Reserva Ancellotta

Other labels: Alto Monte

Visiting: Open to the public; tasting room; restaurant

Located in the small town of Alto Feliz, at the southern entry into Serra Gaúcha, this stunningly attractive winery is seamlessly built into the landscape, with modern architecture and a glass facade overlooking the vineyards. Don Guerino makes a diverse range of top-notch wines, including the best Moscato Giallo I have had outside Italy. With its Ancellotta and Teroldego offerings completing a regional trifecta, Don Guerino stays true to its northern Italian roots.

Don Laurindo

Founded: 1991

Address: Estrada do Vinho, 8 da Graciema, Vale dos Vinhedos, Bento Gonçalves, Rio Grande do Sul, www.donlaurindo.com.br

Owner: Ademir Brandelli

Winemaker: Ademir Brandelli

Viticulturist: Ademir Brandelli

Known for: Tannat, Malbec, Ancellotta, Cabernet Sauvignon, Merlot, Chardonnay, Malvasia de Cândia

Signature wine: Gran Reserva

Other labels: Júlio Brandelli

Visiting: Open to the public

A fourth-generation winemaker, Ademir Brandelli is considered one of Brazil's finest enologists. He cut his teeth at the celebrated Dal Pizzol for several years before returning to the family business. His emphasis on French and Italian varieties reflects a tribute to his roots and his admiration of European classics. The winery is named for Ademir's father, Laurindo. It is situated on a thirty-acre property that has been in the family since 1946; after Laurindo took it over, he focused on *vinifera*. Almost all the winery's sales are direct: retail, restaurant, and online. In 2005 it bottled a Tannat-based red blend to celebrate the tenth anniversary of its first bottling of Tannat, a pioneering feat in Brazil.

Dunamis Vinhos e Vinhedos

Founded: 2011

Address: Rua Rui Barbosa 1319, Dom Pedrito, Rio Grande do Sul, www.dunamisvinhos.com.br

Owner: José Antônio Peterle

Winemaker: Thiago Salvadori Peterle

Viticulturist: Alécio Ciro

Known for: Pinot Grigio, Merlot, Cabernet Franc, Chardonnay, Muscat, Pinot Noir, Tannat, Sauvignon Blanc, Cabernet Sauvignon

Signature wine: Shall We Dance? Cabernet Franc

Visiting: Open to the public by appointment only

Though based in Campanha Gaúcha's Bagé region with thirty-seven acres, Dunamis has two dozen acres of vineyards up north in Serra Gaúcha's Cotiporá. The name comes from Greek and means "power."

Garibaldi

Founded: 1931

Address: Avenida Rio Branco 833, Garibaldi, Rio Grande do Sul, www.vinicolagaribaldi.com.br

Owners: Cooperative with 350 members

Winemaker: Gabriel Caríssime

Viticulturist: Co-op members

Known for: Muscat, Chardonnay, Merlot, Cabernet Sauvignon, Tannat, Riesling, Cabernet Franc, Malvasia

Signature wine: Muscat sparkling wine

Visiting: Open to the public; tasting room

In the 1940s, Garibaldi was the largest cooperative winery in South America. Today it remains formidable and has made a resounding improvement in quality. Located in the eponymous town, Garibaldi is made up of 350 families who collectively farm about two thousand acres. Specializing in sparkling wines, it makes a dizzying seventy-four different offerings, of which about half are fine wine and the rest common table wines. Almost four hundred thousand bottles' worth of wine, originating from the vineyards of ten families, is certified organic.

Geisse

Founded: 1979

Address: Linha Jansen sem número, Pinto Bandeira, Bento Gonçalves, Rio Grande do Sul, www.vinicolageisse.com.br

Owner: Mario Geisse

Winemaker: Mario Geisse

Known for: Chardonnay, Pinot Noir

Signature wine: Cave Geisse Terroir Nature

Other labels: Cave Amadeu, El Sueño

Visiting: Open to the public; tasting room

The importance of Geisse in South America cannot be overestimated: it is arguably the finest producer of sparkling wines on the continent. Its Chilean owner, Mario Geisse, is a driven and focused artist who has worked with Chandon. He and his son Daniel cultivate their grapes in Pinto Bandeira, now an IP for sparkling wine that Geisse discovered in his initial exploration of Brazil for Chandon. They also make some still wine, which is available at the winery, but 85 percent of their offerings are effervescent, and all of those are hand riddled. Sold almost exclusively to restaurants, Geisse wines are a testament to a man with a dream. Geisse also oversees the production at the Chilean winery Casa Silva and has a joint-venture sparkling-wine project in Brazil with Champagne's Philippe Dumont.

Grupo Decima

Founded: 1930

Address: Rua Visconde de Pelotas 2188, Caxias do Sul, Rio Grande do Sul, vinhosdecima.com.br

Owners: Luiz David Travesso, Adriana Maroni Travesso, Alejandro Cardozo, Federico Troentle

Winemaker: Alejandro Cardozo

Known for: Prosecco, Viognier, Tannat, Petit Verdot

Signature wine: X Decima

Other labels: Piagentini, Boutiq Brasil, Cordon d'Or, Piave

Visiting: Open to the public by appointment only; tasting room

Decima was founded eighty years ago when the Piagentini family arrived from Italy. Its two main labels for fine wine are Decima (though few bottles are sold under that label) and Boutiq Brasil. The winemaker, Alejandro Cardoso, is from Uruguay. Julia Harding, Master of Wine, has pointed out that this winery is one of the first to make use of the experimental Champagne practice of using yeast encapsulated in beads (removing the need for *remuage*).

Kranz

Founded: 2008

Address: Rua dos Pioneiros 220, Centro Treze Tílias, Santa Catarina, www.vinicolakranz.com.br

Owner: Walter Melik Kranz

Winemaker: Cleimar Reginatto

Consulting winemaker: Jean Pierre Rosier

Known for: Merlot

Signature wine: Kranz Merlot

Visiting: Open to the public; tasting room

The Kranz winery is one of the most high-tech in South America, as one would expect of an owner who was a longtime executive with Mercedes-Benz. With a history of employing foreign consultants, Austrian and now French, the winery crafts European-style wines in this chillier appellation. Like many wineries in Brazil, the company also has a very successful juice and conserve business. As the winery does not have its own vineyards, it follows two clear rules when negotiating with its suppliers: it accepts only high-quality fruit, and the grapes must come from vineyards at altitudes above 3,900 feet. The winery is part of Kranz's larger business group, which includes Kranz Transportation, Kranz Winery, and Kranz Food Technology.

Laurentia

Founded: 1996

Address: BR 116, km 316, Estrada do Cortado sem número, Barra do Ribeiro, Rio Grande do Sul, www.laurentia.com .br

Owners: Gilberto Schwartsmann, Leonor Schwartsmann

Winemaker: Fausto Filippon

Viticulturist: Evalde Antônio Filippon

Known for: Montepulciano, Carmenère, Tempranillo, Chardonnay, Pinot Noir, Maccabeo, Prosecco, Cabernet Franc, Cabernet Sauvignon, Nebbiolo

Signature wine: Montepulciano

Visiting: Open to the public by appointment only

Outside the classic Serra Gaúcha wine country, Laurentia is situated just south of Porto Alegre. Its location makes it popular with local tourists: it sees close to 100,000 visitors per year and does an extensive wedding business on the beautiful grounds, which also house an art gallery similar to those of the Hess Collection and Clos Pegase in the Napa Valley. Its focus is on Italian varieties—Montepulciano, Sangiovese, and Nebbiolo—and other uncommon grapes, including the Spanish Tempranillo and Albariño. Almost 60 percent of the wines are sold in Rio de Janeiro and direct at the winery.

Lidio Carraro

Founded: 2001

Address: RS 444, km 21, Estrada do Vinho, Vale dos Vinhedos, Bento Gonçalves, Rio Grande do Sul, www.lidiocarraro.com

Owner: Lidio Carraro family

Winemakers: Giovanni Carraro, Juliano Carraro

Consulting winemaker: Monica Rossetti

Viticulturist: Lidio Carraro

Known for: Merlot, Cabernet Sauvignon, Cabernet Franc, Tannat, Touriga Nacional, Malbec, Pinot Noir, Chardonnay, Teroldego, Nebbiolo, Tempranillo

Signature wine: Grande Vindima

Other labels: Singular, Elos, Sul Brasil Agnus, Sul Brasil Dádivas

Visiting: Open to the public; tasting room

A genuine family affair, the Lidio Carraro winery is located adjacent to Miolo (see below). After multiple generations of growing grapes exclusively, the proprietors decided to try to make a go of independent winemaking in the early 2000s. The affable patriarch, Lidio Carraro, is overall director and takes care of the vineyards; his wife, Isabel, is in charge of administration and tourism. Their daughter Patrícia oversees marketing and exports, and their sons Juliano and Giovanni are both winemakers, assisted by Monica Rossetti, who also makes wine in Italy. Carraro's groundbreaking efforts in Encruzilhada do Sul have been crucial for the country as well as the winery. The most novel feature of its winemaking is that it uses no oak whatsoever in order to showcase the purity of the fruit.

Miolo Wine Group

Miolo is the largest privately owned wine entity in the country. Focusing on fine wine, the group has eight different projects, including five in Brazil, two in Italy, and one each in Chile, Spain, Argentina, and France. In Brazil, the wines made in the Vale dos Vinhedos come under the Miolo label, whereas those from Campanha, on the border with Uruguay, are bottled under the Seival label. Miolo owns over 2,800 acres of *vinifera* vines and buys from over five dozen families. It has 40 percent of the domestic market in fine wines and 15 percent of the huge sparkling-wine market. It is also the largest exporter of Brazilian wines. The operation is overseen by Adriano Miolo; two brothers and a sister also participate.

In addition to the Miolo winery (listed below), the company owns the following four Brazilian wineries:

Lovara: Originally the project of Henrique Benedetti, this boutique operation runs autonomously and in the same spirit established by the original owner. It is located in the Vale dos Vinhedos and focuses on red wines.

Ouro Verde: Acquired in 2000 as a partnership between Miolo and Lovara's Henrique Benedetti, this enterprise (known internally as Terra Nova) is located near the banks of the São Francisco River in the municipality of Casa Nova, Bahia.

RAR: A winery located in the Campos de Cima da Serra not far from Vacaría, RAR stands out in an area known more for apples than grapes and quite close to São Joaquim in Santa Catarina just to the north.

Seival: The Quinta do Seival, named for a celebrated battle during Brazil's war of independence, is a crown jewel of the Miolo empire. The property, near Bagé in western Campanha Gaúcha, was first planted in 2001. With 415 acres, the vineyards contain seventeen different varieties. (It used to be nineteen, but they have since pulled up their Alfocheiro and Touriga Franca.)

Miolo Winery

Founded: 1989 (as a group)

Address: RS 444, km 21, Linha Leopoldina, Vale dos Vinhedos, Bento Gonçalves, Rio Grande do Sul, www.miolo.com.br

Owners: Miolo, Randon, and Benedetti families

Winemaker: Adriano Miolo

Viticulturist: Ciro Pavan

Known for: Merlot, Chardonnay, Cabernet Sauvignon

Signature wines: Lot 43, Terroir Merlot, Miolo Reserva, Miolo Seleção, Ouro Verde Testardi Shiraz

Visiting: Open to the public; tasting room; restaurant (Osteria Mamma Miolo)

The Miolo winery is as attractive as it is functional. The main building is a multi-level construction with gardens and a large pagoda, surrounded by grapevines of different varieties. Across the highway is the famous Hotel and Spa Do Vinho, whose owners have partnered with the winery. Through 2012, Miolo worked with Michel Rolland. Its "Lot 43" and "Terroir Merlot" are considered two of Brazil's most consistent iconic wines.

Panceri

Founded: 1990

Address: Linha Leãozinho, Tangará, Santa Catarina, www.panceri.com.br

Owners: Celso Panceri, Luiz Panceri

Winemaker: Marcel Giovani Salante

Viticulturist: Luiz Panceri

Known for: Cabernet Sauvignon, Merlot, Teroldego

Signature wine: Merlot Panceri Reserve

Other labels: Teroldego Nilo, Altos

Visiting: Open to the public; tasting room

In 1884, when Giuseppe Panceri immigrated to Brazil from the Lombardy region in Italy, he brought with him the family tradition of cultivating vineyards, which is now in the fifth generation in Santa Catarina. Thus, like many Brazilian winemakers, the Panceri family were originally growers before establishing their own winery. In 2003, they opened a museum, which documents the 130 years of Italian presence and 80 years of winemaking in Santa Catarina. Renovated in 2009, the museum is important for tourism and as a historical resource for universities.

Pericó Vinhos

Founded: 2001

Address: Rua Valdomiro Pereira do Cruz sem número, Tres Pedrinhas, Santa Catarina, www.vinicolaperico.com.br

Owner: Wandér Weege

Winemaker: Jefferson Sancineto Nunes

Known for: Cabernet Sauvignon, Merlot, Sauvignon Blanc, Pinot Noir

Signature wine: Cave Pericó

Other labels: Vinho do Gelo, Taipa, Vigneto, Plume, Basaltino, Basalto

Visiting: Open to the public by appointment only; tasting room

In June 2009, Pericó claimed the title of creator of the first authentic ice wine in Brazil. The estate was also the first to produce sparkling wines in the region and is considered an innovator. Named for the surrounding valley in São Joaquim, Pericó is located at one of the highest elevations in the country (4,250 feet). It is the culmination of a dream for Wandér Weege, a successful Santa Catarina businessman.

Perini

Founded: 1970

Address: Santos Anjos, Farroupilha, Rio Grande do Sul, www.vinicolaperini.com.br

Owner: Benildo Perini

Winemaker: Leandro Santini

Viticulturist: Marcos Leonardo Lazzari

Consultant: Mario Geisse

Known for: Cabernet Sauvignon, Merlot, Tannat, Chardonnay, Marselan, Muscat, Riesling, Prosecco

Signature wines: Qu4tro, Éden

Other labels: Casa Perini, Arbo, Jota Pe, Pretinha

Visiting: Open to the public by appointment only; tasting room; restaurant (Taverna Perini)

Coming from the Veneto in 1876, the Perinis were one of the founding families of the Forqueta co-op in Farroupilha back in 1920. Fifty years later, frustrated with the quality of the co-op's production, they set out to make their own wines. They began at a very basic level with table wines but have since greatly expanded their range, which now includes super-premium wines. Specialists in Muscat, they are also great proponents of Marselan and make one of Brazil's top examples. Perini has a well thought-out visitor center and tasting room and a good restaurant. Their Éden, a sweet, oak-aged, fortified dessert wine made from aromatic white grapes, is one of the country's best.

Peterlongo

Founded: 1915

Address: Rua Manoel Peterlongo 216, Garibaldi, Rio Grande do Sul, www.peterlongo.com.br

Owners: Luiz Carlos Sella, Adilson Luiz Bohatczuk

Winemaker: Ricardo Morari

Known for: Chardonnay, Pinot Noir

Signature wine: Elegance Champenoise Brut

Other labels: Presence, Privillege, Armando, Terras, Manolo, Must

Visiting: Open to the public

The Peterlongo family came to Rio Grande do Sul over a hundred years ago. Claiming to be the first producers of *méthode traditionnelle* sparkling wine not only in Brazil but on the entire continent, Peterlongo has specialized from the onset in sparkling wines, and it produced nothing else until 1965. The company's facilities include a castle-like building and an underground cellar encased in basalt rock that helps maintain a consistent cellar temperature, ideal for longer-term aging of sparkling wines on their yeast (*en tirage*).

Pizzato

Founded: 1999

Address: Via dos Parreirais sem número, Vale dos Vinhedos, Bento Gonçalves, Rio Grande do Sul, www.pizzato.net

Owner: Plinio Pizzato family

Winemaker: Flavio Pizzato

Viticulturist: Plinio Pizzato

Known for: Merlot, Chardonnay, Tannat, Cabernet Sauvignon, Alicante Bouschet, Egiodola, Pinot Noir

Signature wine: DNA99 Merlot

Other labels: Fausto

Visiting: Open to the public; tasting room

In a short time, Flavio Pizzato has built his small family winery into one of the most celebrated in Brazil. Plinio Pizzato, Flavio's father, is in charge of the vineyards, most of which are in the heart of Vale dos Vinhedos. In the early years, the family focused on Merlot while they worked on dramatically improving the vineyards. Over time they increased their offerings, passionately promoting more obscure varieties like Egiodola, Alicante Bouschet, and Marselan. Flavio's wife, Giovanna, is an exceptional chef, and the winery offers eight-course tasting meals showcasing their Italian heritage and Brazilian specialties presented with modern flair.

Quinta Santa Maria

Founded: 2004

Address: Manoel Joaquim Pinto 348, Centro, São Joaquim, Santa Catarina, www.quintasm.com

Owner: Antonio Nazário dos Santos

Winemaker: Jean Pierre Rosier

Known for: Touriga Franca, Touriga Nacional, Tinta Roriz, Muscat, Trebbiano

Signature wines: Grand Utopia, Portento Tinto

Other labels: QSM Blend

Visiting: Open to the public; tasting room

The idea of producing wines with Portuguese spirit is the driving philosophy of the Portuguese owner Antonio Nazário dos Santos and his seven partners. Situated in a scenic location replete with araucarias and apple trees, Quinta Santa Maria was envisioned as a miniature version of Portugal's Douro Valley. The winery's elevation and the abundance of Portuguese varieties make you understand why. The owner is both passionate and irreverent: he makes a cherry liqueur and a distinctive Cabernet Sauvignon–inspired, Amarone-like wine.

RAR. *See* Miolo Wine Group

Routhier & Darricarrère

Founded: 2008

Address: BR 158, km 473, Rosário do Sul, Rio Grande do Sul, www.redvin.com.br

Owners: Pierre Darricarrère, Jean Daniel Darricarrère, Michel Routhier

Winemaker: Anthony Comerlatto Darricarrère

Viticulturist: Jean Daniel Darricarrère

Known for: Cabernet Sauvignon, Chardonnay, Merlot

Signature wine: Província de São Pedro

Other labels: RED

Visiting: Open to the public; tasting room

This postage stamp–sized operation is a partnership of two families that have been friends for over thirty years and previously worked together in the citrus business. The Routhiers are French Canadian, and the Darricarrères are longtime South Americans of African *pied-noir* descent. The winery was Anthony's inspiration, and wine is in his blood: his great-grandfather was a winemaker in Algeria, and he is a cousin of the late Anthony Perrin of Bordeaux's Château Carbonnieux and Genevieve Janssens, director of winemaking at Robert Mondavi in the Napa Valley. The minimal-intervention approach applied here includes the use of native yeasts and minimal filtration.

Salton

Founded: 2008

Address: Rua Mário Salton 300, Distrito de Tuiuti, Bento Gonçalves, Rio Grande do Sul, www.salton.com.br

Owner: Salton family

Winemaker: Lucindo Copat

Known for: Cabernet Sauvignon, Chardonnay, Merlot, Cabernet Franc, Tannat, Pinot Noir, Riesling

Signature wine: Talento

Other labels: Desejo, Gerações, Virtude, Poética, Volpi, Intenso, Lunae, Classic, Flowers, Serra Nevada, Chalise, Perlage

Visiting: Open to the public; tasting room; restaurant

One of Brazil's oldest and largest wineries is still family owned and run: Daniel Salton, the current president and director, is a third-generation Salton. Tuiuti, the site of the new Vale do Rio das Antas winery, is actually where the Saltons began before moving to Bento: when they outgrew the Bento facility, they returned home. Seven hundred families provide 60 percent of Salton's fruit; the remaining 40 percent comes from their own vineyards. The winery makes a range of close to seventy wines, ranging from everyday table wines to some of the country's best bottles. For the past six years it has been the country's leading seller of sparkling wine. Salton is the first Brazilian winery to be fair-trade certified.

Sanjo

Founded: 1993

Address: Av. Irineu Bornhausen 677, São Joaquim, Santa Catarina, www.sanjo.com.br

Owner: Cooperative

Consulting winemakers: Anderson Schmitz, Marcos Vian

Known for: Cabernet Sauvignon

Signature wine: Maestrale Cabernet Sauvignon

Other labels: Nobrese, Núbio

Visiting: Open to the public; tasting room

Sanjo is a small and unusual Brazilian co-op: 90 percent of the company's founders are Japanese immigrants or of Japanese descent. After the Cotía co-op in São Paulo state folded, these families moved en masse to Santa Catarina and were employed in the apple industry, of which São Joaquim lies at the heart both geographically and commercially. Sanjo is a leading player in the country's market for apple juice and related products. In 2002, seeking to diversify, it began cultivating grapes. With seventy-four acres under vine, and despite the cool climate, the winery is focused on Cabernet Sauvignon.

Suzin

Founded: 2007

Address: Rua Juiz Fonseca Nunes 379, São Joaquim, Santa Catarina, www.vinicolasuzin.com.br

Owners: Everson Fernando Suzin, Jeferson Luiz Suzin, Zelindo Melci Suzin

Consulting winemaker: André Tonet

Viticulturists: Everson Fernando Suzin, Jeferson Luiz Suzin, Zelindo Melci Suzin

Known for: Cabernet Sauvignon

Signature wine: Suzin Cabernet Sauvignon

Visiting: Open to the public by appointment only; tasting room

André Tonet is the talented winemaker who oversees the production at Suzin and of his own label, Domans. The winery sells off about 60 percent of its fruit and keeps only the best lots. It is a member of Acavitis (Associação Catarinense dos Produtores de Vinhos Finos), an association of close to thirty producers in Santa Catarina with a commitment to quality and *vinifera* wines (www.acavitis.com.br).

Villa Francioni

Founded: 2000

Address: Rodovia SC 438, km 70, São
Joaquim, Santa Catarina,
www.villafrancioni.com.br

Owners: Manoel Dilor Francioni, Adriana
Freitas, André Freitas, Daniela Freitas,
João Paulo Freitas

Winemaker: Organlindo Bettú

Consulting winemaker: Gustavo Gonzalez

Viticulturist: Organlindo Bettú

Consulting viticulturalist: Michael Raymor

Known for: Chardonnay, Sauvignon Blanc,
Cabernet Sauvignon, Merlot, Cabernet
Franc, Sangiovese, Pinot Noir, Petit
Verdot, Malbec, Syrah

Signature wine: Michelli

Other labels: Aparados, Joaquim

Visiting: Open to the public by appointment only; tasting room

The first thing you notice at this winery is the winery building itself. Stained-glass windows are incorporated into a six-story structure housing the production facility, a restaurant, an art gallery, and shops. No detail in design has been overlooked, from the fanciful wrought-iron balconies overlooking the storage tanks to the mosaic tile floors under the aging barrels. The gallery has exhibited the works of artists such as the internationally celebrated Camille Claudel and Brazil's Luciano Martins. In 2003, the owner, Manoel Dilor Francioni, visited the Robert Mondavi Winery. Inspired by Mondavi's role as a leader in the American wine industry, Francioni dedicated himself to making his own winery a leader in Brazil. Though he passed away before realizing his vision, his children have continued the effort.

Villaggio Grando Boutique Winery

Founded: 1998

Address: Rodovia SC 451, km 56,
Água Doce, Santa Catarina,
www.villaggiogrando.com.br

Owner: Maurício Carlos Grando

Winemakers: Mateus Valduga, Antonio
Saramago, Jean Pierre Rosier

Known for: Chardonnay, Petit Manseng,
Gros Manseng, Malbec, Cabernet Franc

Signature wines: Innominabile, Além Mar

Other labels: Chance, Colheita Tardia

Visiting: Open to the public; tasting room

Situated in Herciliópolis in Água Doce, Villaggio Grando was established after two French producers, one from Armagnac and one from Bordeaux, met with Maurício Carlos Grando and encouraged him to take the plunge into winemaking. Their confidence was justified. The winery has over a hundred varieties planted on 104 acres, ranging from the expected (Chardonnay) to the unexpected (Petit Manseng). It plans to begin distillation and introduce a brandy soon, presumably a tribute to the man from Armagnac who encouraged the project.

Vinibrasil (Vitivinícola Santa Maria)

Founded: 2002

Address: Estrada da Uva e do Vinho, km 8, Lagoa Grande, Pernambuco, www.vinibrasil.com.br

Owner: Dão Sul

Winemakers: Osvaldo Amado, Ricardo Henriques

Viticulturist: Rogério de Castro

Known for: Cabernet Sauvignon, Touriga Nacional, Aragonês, Alicante Bouschet, Syrah, Viognier, Chenin Blanc, Fernão Pires, Moscato Canelli

Signature wine: Rio Sol

Other labels: Adega do Vale, Vinha Maria, Paralelo 8, Rendeiras, Allure, Matuto, Blisse

Visiting: Open to the public; tasting room; restaurant

The winery is owned by the leading Portuguese company Dão Sul, whose reputation for quality has redefined the reputation of the wines of the Beiras region. São Francisco is known more for palm trees and coconuts than for quality wine, but this project aspires to defy that stereotype.

At this site, consisting of almost five hundred acres along the São Francisco River, Dão Sul spent more than $4 million to set up the facility with state-of-the-art equipment. Vinibrasil produces more than two dozen different wines.

Vinícola Santa Augusta

Founded: 2006

Address: Linha Santa Barbara, Videira, Santa Catarina, www.santaaugusta .com.br

Owner: Taline de Nardi

Winemaker: Cássia Savi

Consulting winemaker: Jefferson Sancineto Nunes

Viticulturists: Alessandro Paviani, Jefferson Sancineto Nunes

Known for: Merlot, Chardonnay

Signature wine: Santa Augusta

Other labels: Sarau

Visiting: Open to the public by appointment only; tasting room; restaurant

With vineyards in Videira and Água Doce, Santa Augusta is one of the few wineries that truly embraces biodynamic farming, to which it committed itself in 2011. It is especially known for its Moscato Giallo *passito* wine (a style of wine made with sun-dried grapes), which has received a number of awards. Santa Augusta is unusual for making only *vinifera* wines.

6

URUGUAY

A grape picker at Pizzorno winery in Canelones.
Photo by Mariano Herrera.

Planted acreage: 22,250

World rank in acreage: 57

Number of wineries: 270

Per capita consumption (liters)/world
ranking: 26.1/17

Leading white grapes: Chardonnay, Sauvignon Blanc, Ugni Blanc, Viognier

Leading red grapes: Black Hamburg,
Cabernet Franc, Cabernet Sauvignon,
Concord, Merlot, Muscat, Tannat

Memorable recent vintages: 2000,
2001, 2002, 2004, 2011, 2012

INTRODUCTION

Uruguay means "land of the painted bird" in the native Guarani language. This diminu-
tive country—covering less than 400 miles north to south and 320 miles from east to
west, with an area fifty times smaller than Brazil's—is home to 450 species of birds, just
50 fewer than are found in the entire Amazon River basin. This biodiversity is echoed
by a surprising diversity in wine.

One of the best-kept secrets in South America, with a national population less than that of the extended San Francisco Bay Area, Uruguay has no car traffic to speak of, a thriving middle class, zero visible homelessness, and a great vibe. It produces some of the tastiest beef I've ever eaten (and I've had my fair share over the years), which is responsible for over one-third of Uruguay's export revenue. It is home to friendly people, a perennially competitive national soccer team, and Punta del Este, the South American equivalent of France's Saint-Tropez.

Almost 90 percent of Uruguay's people are of European descent, mainly Spanish and Italian; there are few indigenous Uruguayans. The original inhabitants, the Charrua, were absorbed into the Spanish and Portuguese populations after a long resistance to colonization. The mestizo contingent (less than 10 percent of the total population) is found principally in northern Uruguay. Almost 50 percent of the country's inhabitants live in the country's capital, Montevideo.

Situated on the Atlantic Coast between Argentina and Brazil, Uruguay has a decidedly maritime climate. With the Río de la Plata and Atlantic Ocean to the south and southeast, the coastal areas are humid but not particularly hot: winds from the Antarctic create moderate temperatures and generally cooler evenings. Farther north, closer to Brazil, the climate can be quite warm, especially in Artigas and Rivera. Eighty-five percent of Uruguay's land is dedicated to agriculture—the highest proportion of any country in the world. Its terrain marks the transition zone between the humid Argentine pampas and the uplands of southern Brazil. North of the alluvial plain, known as the Banda Oriental, are long, sweeping slopes and grasslands, meandering rivers, and long ranges of low hills, rarely exceeding one thousand feet above sea level. Climatic variations are moderate, and rainfall is evenly distributed through the seasons.

Although Uruguay is the fourth largest wine producer in South America, it's all relative. The entirety of Uruguay's annual output (approximately 68 million cases) is less than that of the Chilean Concha y Toro group. Vineyards are concentrated in the coastal regions of greater Montevideo, San José, and Canelones, all within an hour of the capital. Canelones alone is responsible for 60 percent of all wine made. As in neighboring Brazil, Uruguay's grape production leans heavily toward American native and hybrid grapes and lesser *vinifera* (such as Ugni Blanc), used for wine, juice, table grapes, and concentrates. Roughly 60 percent of the wineries make exclusively "common" wine, much of which is sold in traditional demijohns. Of the country's 270 wineries, over 90 percent are family owned, with the typical operation vinifying around twenty thousand cases—20 percent bottled quality wine and 80 percent "commún."

Only three dozen or so wineries are focused on export-quality fine wines. There have been three distinctive waves of entrants into Uruguay's wine trade: the "originals," including Carrau, Ariano, Bodegas Castillo Viejo, Pisano, and Stagnari; a second wave containing Irurtia, Filgueira, Marichal, and Pizzorno; and the "new arrivals," including Bouza, Finca Narbona, Alto de la Ballena, Bodega Garzón, and Viñedo de los Vientos. Foreign consultants in Uruguay are few: they include Michel Rolland at Narbona,

Alberto Antonini at Garzón, and Paul Hobbs at Juanicó. The South American "resident Kiwi" Duncan Killiner consults for Alto de la Ballena, Juanicó, and Pizzorno.

Today, Uruguayans drink about 70 percent domestic wines (lower than official esti-mates) and about 30 percent imports. Fifteen years ago the balance was closer to 50/50, and the proportion of domestic wine is steadily increasing, although overall wine con-sumption has declined.

HISTORY

In 1870, a Basque settler named Pascual Harriague planted 490 acres of Tannat in the northern region of Salto along the Uruguay River. Now the signature variety of the country, Tannat was locally named Harriague in his honor. While Harriague was planting in Salto, a Catalan named Francisco Vidiella was cultivating Folle Noir grapes in Colón, closer to Montevideo. Folle Noir was subsequently renamed Vidiella. Of infe-rior quality, it has disappeared over time. Other grapes were gradually added (mostly Spanish and Italian varieties), and in 1903, recognizing the success of Uruguay's wine industry, the government passed a law governing viticulture. The industry thrived. As in Brazil, eager regional merchants planted vast quantities of highly adaptable Ameri-can *labrusca* varieties, with plantings reaching 47,000 acres by the 1950s.

Many grapes were simply and deferentially called Harriague and Vidiella until the 1970s. In 1976, as French consultants were hired in an effort to improve quality, Denis Boubais from Montpellier identified Harriague as Tannat and Vidiella as Folle Noir. Domestic consumption expanded between the 1950s and the 1980s, when the Merco-sur, the trade and political alliance among Brazil, Argentina, and Paraguay (with Chile as an associate member), was established.

Anticipating the arrival of free trade and detecting signs of increasing wine consumption, Uruguayan producers saw that a more open market for wine would give them only two options: cede the entire market to Chile and Argentina, or prepare themselves to compete. The decision to improve resulted in what is now known as "the reconversion" of the 1980s, when the government assisted the wineries in replanting for quality. Led by the grower Reinaldo De Lucca, who went to Montpellier to study, winemakers imported better clones, starting with the iconic Tannat. De Lucca's nursery is still a significant reservoir of Uruguayan grape diversity, and De Lucca remains the largest importer of plant material.

At the zenith of Uruguay's winemaking industry, there were 877 wine companies. By 1990, there were 426, and today the number is around 270. Some of this decline is due to consolidation, but many wineries have simply closed shop as consumers have moved away from common wines to fine wines and other beverages. When asked about the dramatic decrease in number of wineries, veteran Edgardo Etcheverry of Castillo Viejo shrugged: "Uruguayans are stubborn, and most would rather fold than merge."

URUGUAYAN GAME CHANGERS

FRANCISCO CARRAU

The passionately engaged Francisco Carrau oversees one of Uruguay's truly dynastic family wineries, dating back to the mid-1700s. He is in charge of winemaking, but to say that he's just a winemaker would be flip. He has a PhD in chemistry, is the professor of enology at the country's one wine school (the Escuela de Vitivinicultura Tomás Berreta), and is responsible for educating the future generations of the trade. Among

MAPPING YOUR GRAPES

The mapping of the human genome in the year 2000 paved the way for genomic studies of many other organisms, not least the wine grape. The first attempt, in 2007, was carried out by the French for Pinot Noir in Burgundy. Although it was not entirely successful, it did reveal that Pinot Noir has about thirty thousand genes in its DNA, some five thousand more than the human genome. But the 2012 announcement of the near-perfect and exponentially less expensive sequencing of Tannat by Francisco Carrau was groundbreaking in its accuracy and its broader applicability to understanding genetic variations in clones and selections, the practical implications of these differences, and how to best take advantage of them in planting.

other efforts, Francisco conducted pioneering studies on native yeasts and their effect on wine quality. More recently he led research on sequencing the genome of Tannat that is considered groundbreaking for global viticulture. Francisco and his brother Juan, an equally gifted viticulturist who is responsible for developing the Cerro Chapeu region in Rivera, are featured in Christopher Fielden's book celebrating the Carrau family, titled *Yet Another Road to Cross*.

FERNANDO DEICAS

Fernando may be the Robert Mondavi of Uruguay. Now the proprietor of Establecimiento Juanicó, the country's largest producer of fine wine, Deicas took over the family bodega in the early 1980s and adopted an innovative approach. In the 1990s he was the first to implement large-scale green harvesting and to put on official sommelier tastings and training. He provided the first decanters to restaurants and established the scarily successful Don Pascual brand, which sells one of every five bottles of fine wine consumed in Uruguay. He has a joint venture with France's entrepreneurial Bernard Magrez, and he works with different consultants on his different lines of wines—from Paul Hobbs to Peter Bright to Duncan Killiner—in a constant effort to improve.

REINALDO DE LUCCA

Reinaldo De Lucca is considered a leader in Uruguayan viticulture, heir to the mantle of Luis Hidalgo, the Spaniard credited with bringing viticultural study and mapping to Uruguay. As mentioned above, Reinaldo traveled to France in the 1980s in search of new vines to import, notably Tannat, and his nursery is the leading source of plant material for the country. He always says what he thinks, occasionally making his colleagues uncomfortable with his bracingly non-PC opinions. One part "Doctor Dirt" and one part pragmatic philosopher, he is perennially consulted on what to plant and where to plant it. He teaches viticulture at the university, and his thesis paper on growing Tannat in different soil types is renowned.

DANIEL PISANO

Daniel Pisano is unforgettable. Proud, enthusiastic, and the leading representative of Uruguayan wine to the export world, Daniel has taken his family's high-quality wines and made them into a global success: Pisano wines can be found in forty-six countries on five continents. A remarkably articulate and knowledgeable spokesperson for the industry, Daniel is equally inventive and creative in marketing and branding. With his brothers Eduardo and Gustavo, respectively responsible for vineyards and winemaking, he has formed an amazing team. He even inspired his nephew Gabriel, now a leading *garagiste* with his own Viña Progreso boutique, to give it a go.

REGIONS AND WINERIES

Uruguay's wine area is close to the water, with most of the vineyards being within a short distance of either the Río de la Plata, beginning at the western border and running east past Montevideo, or the Atlantic Ocean, extending north from Punta del Este to the Brazilian border at Chui. Only 22,250 acres are cultivated. Although grapes are grown in sixteen of the country's nineteen administrative departments, four departments account for 90 percent of plantings.

White varieties make up about 20 percent of Uruguay's fine-wine grapes, led by Chardonnay and Sauvignon Blanc and followed by Viognier and Muscat Ottonel. Of the red varieties, close to 60 percent of the acreage is in Tannat, followed by Merlot at 20 percent and Cabernet Sauvignon at 15 percent, with the remaining 5 percent being a mixture of different grapes. Tannat is not only historically important but also amazingly good, with different regions having distinctive Tannat flavors, as with Malbec in different parts of Argentina. The key to improving Tannat has been a better understanding of the grape and its tannin management. This understanding began in the mid-2000s, when producers learned to leave the fruit hanging as long as possible and not to be afraid of some shriveling or raisining. Equally important has been an awareness of the timing of the vine's flowering and fruit-set periods, employing tactical leaf plucking at those precise moments. Managing tannins in the vineyards by substantially dropping yields (to maximize fruit while balancing tannins and herbaceousness) and then picking as late as possible for full lignification of seeds and stems has proved highly successful. Finally, in the winery, contrary to conventional wisdom, prefermentation cold maceration, rather than micro-oxygenation, is being widely adopted to soften the tannins in the wines.

While there are discernible Uruguayan terroirs, and areas known for producing

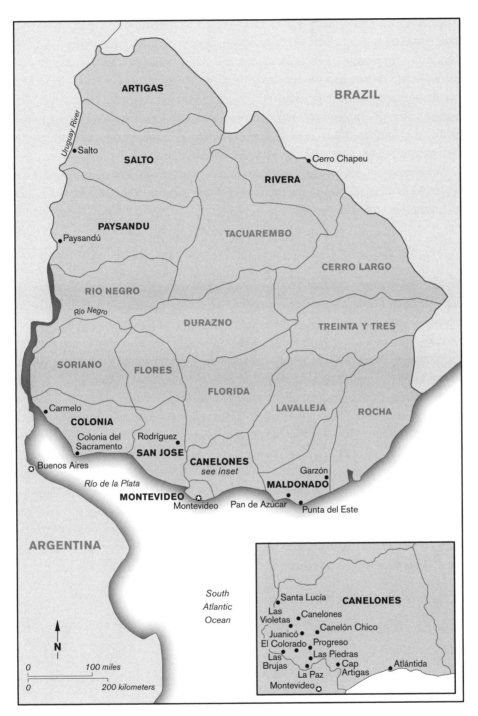

Map 9. Uruguay's wine regions

superior wines, establishing appellations would be premature. There is, however, a label designation for quality: *vino de calidad preferente*, or VCP. To qualify for this label, a wine must be made entirely of *vinifera* grapes, meet a minimum alcohol level, and be bottled in 750 ml bottles with a registered seal (with a bottle number issued by INAVI, the Uruguayan wine institute). Varietally labeled wines must contain a minimum of 85 percent of the specified grape. However, there are no means of establishing subjective quality standards for the wines, such as sampling by tasting panels. Changes to the system have been proposed but not yet implemented.

Most grapes planted are still American *labrusca*. Reflecting the increasing sophistication of the local palate, however, a 2008 law prohibits planting of any new *labrusca* vines (though enforcement is spotty).

NORTHERN AREAS: ARTIGAS, PAYSANDÚ, RIVERA, AND SALTO

Artigas, the northwesternmost department, bordering the Uruguay River, is decidedly hot and sandy and not renowned for quality wine. It is remarkable mainly for being the point where Uruguay, Brazil, and Argentina meet. Paysandú, also in the west, is an industrial center that some believe has potential for wine. It also has substantial agriculture, with large cattle and sheep ranches in addition to farms that grow soybeans, wheat, barley, oranges, and, more recently, blueberries. Constancia, considered the best subregion for grapes, is located nine miles north of the city of Paysandú.

Salto is situated on the right bank of the Uruguay River, across from Argentina's metropolis of Concordia. Now Uruguay's second-largest city (after Montevideo), it is the terminus for the shallow-draft vessels that dredge the river. It is the symbolic birthplace of Uruguay's wine industry, and notably the site of Pascual Harriague's plantings of Tannat. The harsh climate of these northern regions and their distance from Montevideo mean that they produce less than 5 percent of the country's wine.

The department of Rivera, situated east of Salto and Artigas, resembles Brazil's Campanha Gaúcha. The Cerro Chapeu area, while remote, is alight with the same excitement as Brazil's viticulture just north of the border.

RECOMMENDED PRODUCERS

Bertolini & Broglio

Bodegas Carrau

Leonardo Falcone

COLONIA

Colonia sits across the Río de la Plata from Argentina. Though they lagged behind the Portuguese, it was the Spanish missionaries who established vines here, among the country's oldest, and bestowed on it the name Colonia. Easily accessible from Buenos Aires by ferry, the town of Colonia del Sacramento boasts a single winery in the namesake town: Los Cerros de San Juan.

Located forty miles from Colonia del Sacramento, the department's largest and most significant appellation is Carmelo, an area recognized for the high limestone content of its soils. The two most significant wineries are Irurtia and Narbona; there are several more small artisanal wineries in the immediate area, most of which make wine by having grapes custom crushed at other facilities. Located slightly farther north than the main wine region of Canelones and a mere half-mile from the water, Colonia has slightly higher temperatures and earlier fruit ripening, but harvest is still typically ten days to two weeks later than elsewhere in Canelones, as growers leave the fruit to ripen fully so that the tannins polymerize, becoming smoother and mellower. Most of the Colonia region was established by *porteños*, Argentineans from Buenos Aires, who settled in the city of Colonia and in Carmelo, which is also home to immigrants from Switzerland and a host of Central European nations whose influence you can trace in the local cheeses and breads.

Colonia's plantings are approximately 70 percent red, 20 percent white, and 10 percent Rosado Moscatel (one of the most widely planted criollas), with fine wine accounting for about 30 percent of total cultivation. Premium wines are characterized by their vibrancy and structure. Colonia is an area of historical significance, with a resurgent wine industry led by Narbona and the custom-crushing boutique wineries.

RECOMMENDED PRODUCERS

Familia Irurtia

Finca Narbona

Los Cerros de San Juan

SAN JOSÉ

To the east of Colonia on the way to Canelones is San José. Wines here develop richness and depth from the darker soils. Castillo Viejo's vineyard in Concordia, near Rodríguez, dates back to the 1800s and is among the oldest commercial vineyards in the country. With a seemingly perpetual breeze that can at times be a downright biting wind, San José pulls in cold air from the Río de la Plata and the San José and Santa Lucía rivers.

CANELONES

Unquestionably the country's most important region, accounting for 60 percent of all production, Canelones surrounds Montevideo (both the city and the suburban wine region). The people residing in Canelones are called *canarios*, as many of them are descended from immigrants from the Canary Islands. The region has soil rich in clay and a maritime climate. Located at 35° south, level with Argentina's Mendoza, and with elevations ranging from three hundred to five hundred feet, it is unofficially divided into more than a dozen subregions.

The main areas include Progreso, the epicenter of the region, which produces excellent and robust wines from land historically planted to wheat. It is a picturesque region, with low hills, and its climate is characterized by sea breezes from the Antarctic. Las Violetas, next to Progreso, is stylistically similar and contains a noted subregion called Echevarría. Juanicó, located next to the town of Canelones proper, has slightly more calcareous soils, while nearby La Paz is rockier and renowned for its pink granite, gravelly terroir, and resulting deep wines. Las Piedras, north of La Paz, is akin to it stylistically, and new vineyards are being established in Cuatro Piedras, a nearby clay-rich area. Santa Lucía, on the river of the same name, is known for its approachable, softer wines. Las Brujas is located west of La Paz and Progreso. El Colorado, the highest ridge in Canelones, is distinguished by a unique heavy clay and limestone terroir, which produces fuller-bodied wines. Canelón Chico is a very well-thought-of subdistrict between Las Piedras and Las Violetas. Cap Artigas, an eastern suburb of Montevideo (not to be confused with the department of Artigas in the northwest), and Atlántida, an eastern zone on the way to Maldonado, near pine forests and the lovely beach town of the same name, round out the communes of Canelones.

RECOMMENDED PRODUCERS

Antigua Bodega Stagnari	Juanicó
Ariano Hermanos	Los Nadies
Artesana	Marichal
Bodegas Carrau	Pagos de Atlántida
Bodegas Castillo Viejo	Pisano
Bouza	Pizzorno
Bracco Bosca	Santa Rosa
De Lucca	Toscanini
Filgueira	Viña Progreso
Giménez Méndez	Viña Varela Zarranz
Grupo Traversa	Viñedo de los Vientos
H. Stagnari	

MALDONADO

Two hours by car due east from central Canelones is the region of Maldonado. The province is better known for the beach resort of Punta del Este than for wine, and there are only a few bodegas. Breathtakingly beautiful, the area is characterized by soils rich in decomposed volcanic rock and constant, cool Atlantic breezes. It has the highest altitudes in all of Uruguay, at about one thousand feet, not far from Garzón, which is both an area and the name of a winery. The Sierra de la Ballena, where the Alto de la Ballena winery is situated, is another subdistrict with granitic soils and a steep slope reminiscent of the Rhône Valley's Côte-Rôtie. The remaining two regions are Sierra de los Caracoles and Pan de Azúcar (west of Sierra de la Ballena). Maldonado is the home of the best Sauvignon Blanc– and Viognier-based wines that I tasted in Uruguay, and it would not surprise me to see winery numbers increase quickly.

RECOMMENDED PRODUCERS

Alto de la Ballena

Bodega Garzón

WINERY PROFILES

A tractor at the Los Cerros de San Juan winery in Colonia. Photo by Mariano Herrera.

Alto de la Ballena

Founded: 2000

Address: Ruta 12, km 16.4, Sierra de la Ballena, Maldonado, www.altodelaballena.com

Owners: Paula Pivel, Álvaro Lorenzo

Winemaker: Soledad Mello

Consulting winemaker: Duncan Killiner

Known for: Merlot, Cabernet Franc, Tannat, Syrah, Viognier

Signature wines: Tannat Viognier, Cetus Syrah

Visiting: Open to the public by appointment only; tasting room

Set in a spectacularly beautiful location, sitting inland and above the Punta de la

Ballena in the Sierra de la Ballena (hence the winery name of Alto de la Ballena, meaning "heights of the whales"), this new winery is a labor of passion and pioneering commitment by the husband-and-wife team of Álvaro Lorenzo, a real-estate veteran and business executive, and Paula Pivel, a former bank executive. They pooled their savings and ventured east to Maldonado from Montevideo at a time when there were just three vineyards of consequence in the region. Their bulldozed vineyards of quartz and granitic soils and the potential to make serious red wines attracted their consultant, Duncan Killiner, who also works with Pizzorno.

Antigua Bodega Stagnari

Founded: 1928
Address: Ruta 5, km 20, Santos
    Lugares, La Paz, Canelones,
    www.antiguabodegastagnari.com.uy
Owner: Virginia Stagnari
Winemakers: Mariana Meneguzzi
    Stagnari, Laura Casell
Viticulturists: Hector Nelson Stagnari,
    Carlo Meneguzzi Stagnari
Known for: Tannat
Signature wines: Osiris Reserva
    Tannat, Prima Donna Crianza en Roble
    Tannat
Other labels: Del Pedregal, Del Pedregal
    Sutil, Prima Donna, Mburucuyá, Osiris,
    Cantharus, Grys, Ragazza
Visiting: Open to the public by appoint-
    ment only; tasting room; restaurant

The Stagnari family arrived in Uruguay from Genoa in 1928. Both sides of the family were from wine backgrounds, and they settled in to grow grapes and produce wines in La Paz. The bodega is uniquely managed by women: the winery's director, Virginia Stagnari; her daughter Mariana, who is a winemaker; and the senior winemaker, Laura Casell, arguably the matriarch of female winemakers in Uruguay and the lone woman graduate in enology of her generation.

Ariano Hermanos

Founded: 1929
Address: Ruta 48, km 15, Las Piedras,
    Canelones, www.arianohermanos.com
Owner: Ariano Hnos SA
Winemaker: Daniel Vlah
Viticulturists: Edgardo Ariano, Sebastián
    Ariano
Known for: Tannat, Cabernet Franc
Signature wine: Don Adelio Ariano
    Tannat
Other labels: Tacuabé, Viña Constancia
Visiting: Open to the public by appoint-
    ment only; tasting room

When they emigrated from Piedmont in 1927, Adelio Ariano and his brother Amilcar came to South America to work for Cinzano in Argentina. Being entrepreneurial, Adelio and his brother soon left to seek their fortunes in Uruguay. Today, now in the third and fourth generations, the Ariano family owns multiple vineyards across the country, from Constancia in Paysandú to various properties in Canelones. In addition to their perennially consistent table wines, they make a delightful *licor de tannat*.

Artesana

Founded: 2007

Address: Ruta 48, km 3.600, Las Brujas,
Canelones, www.artesanawinery.com

Owner: Blake Heinemann

Winemakers: Analía Lazaneo, Valentina
Gatti

Viticulturist: Eduardo Felix

Known for: Tannat, Merlot, Zinfandel

Signature wine: Tannat

Visiting: Open to the public by appointment only; tasting room

Blake Heinemann, from Malibu, California, came to Uruguay in the early 2000s,
fell in love with the country, and bought
land in Las Brujas, where he established a
winery in 2007. On this eighty-acre estate
winery focusing exclusively on red wines,
two young but experienced and dynamic
female winemakers oversee Artesana's
production. In a tip of the hat to its American parentage, the winery is the only producer of Zinfandel in the country.

Bertolini & Broglio

Founded: 2006

Address: Ruta 3, km 468, Paysandú,
(no website)

Owners: Luis Bertolini, Juan Pedro Broglio

Winemakers: Juan Pedro Broglio, Edgar
Barrera, Walter Barrera

Viticulturist: Juan Pedro Broglio

Known for: Tannat

Signature wine: Tannat Premium

Visiting: Open to the public; tasting room

Located eleven miles south of the town
of Salto, Bertolini & Broglio is a very new
winery in an area that, although historically significant, has until recently not
been known for quality wine. Luis Bertolini and Juan Pedro Broglio, the latter a
noted agronomist, decided to change that
perception. Bertolini, a lawyer by profes-
sion, is savvy about both vineyard management and the Uruguayan wine industry despite having no professional training
in winemaking or viticulture. The winery
keeps only the top 30 percent of its crop,
selling off the rest on the open market; it
is known for its concentrated reds.

Bodega Garzón

Founded: 2009

Address: Ruta 9, km 175.5, Paraje Garzón,
Maldonado, www.bodegagarzon.com

Owner: Alejandro Bulgheroni

Winemaker: Germán Bruzzone

Consulting winemaker: Alberto Antonini

Viticulturist: Eduardo Felix

Known for: Tannat, Albariño, Viognier

Signature wine: Garzón Tannat

Visiting: Open to the public by appointment only; tasting room; restaurant

This extraordinary 9,880-acre property
is situated about forty-five minutes outside Punta del Este. It is divided into two
businesses: Colinas de Garzón (olive oil)
and Bodega Garzón (wine), both owned
by the Argentinean businessman Alejandro Bulgheroni. Colinas de Garzón has already won numerous awards for its varietal olive oil in Argentina and Italy. The
winery employs the renowned Alberto
Antonini as its consulting winemaker. In
addition to wine tasting, it offers visitors

tractor rides and horse-drawn carriage rides around the property, an interactive demonstration in the forty-seat auditorium, and tastings of wine and olive oil with tapas—along with biking, hot-air balloon rides, walking paths, private lunches, picnic grounds, and the opportunity to adopt an olive tree (and enjoy its bounty).

Bodegas Carrau

Founded: 1930

Address: Ruta César Mayo Gutiérrez 2556, Montevideo, www.bodegascarrau.com

Owner: Carrau family

Winemaker: Francisco Carrau

Consulting winemaker: Octavio Gioia

Viticulturist: Javier Carrau

Known for: Tannat, Sauvignon Blanc, Pinot Noir, Petit Verdot, Chardonnay, Petit Manseng

Signature wine: Amat

Other labels: Castel Pujol, Cepas Nobles, Ysern, Xacrat, Sust, Vilasar, Vivent, Arerunguá

Visiting: Open to the public by appointment only; tasting room; restaurant

Few families in the world can boast to have maintained a tradition of winemaking for nine generations, but the Carrau family is one of them. In April 1752, Francisco Carrau Vehils acquired the family's first vineyard in the region of Vilasar de Mar, Catalonia. All subsequent generations of the family have stayed the course. In 1975, Juan Pujol Carrau founded a winery in Las Violetas and launched Castel Pujol, one of Uruguay's first fine wines. Around the same time, the Carraus ventured north and started their celebrated

vineyard in the Cerro Chapeu area of Rivera; a Carrau winery across the border in Brazil opened in 2013. Today, Juan Carrau's five grandchildren continue the tradition. The history of the Carrau dynasty is explored in Christopher Fielden's book *Yet Another Road to Cross.*

Bodegas Castillo Viejo

Founded: 1927

Address: Ruta 68, km 24, Las Piedras, Canelones, www.castilloviejo.com

Owners: Edgardo Etcheverry, Ana Etcheverry, Alejandro Etcheverry

Winemaker: Alejandro Etcheverry

Consulting winemaker: Patrick Soye

Viticulturists: Edgardo Etcheverry, Ana Etcheverry, Alejandro Etcheverry

Known for: Tannat

Signature wine: CataMayor El Preciado

Other labels: Vieja Parcela, Isla de Lobos

Visiting: Open to the public by appointment only; tasting room; restaurant

The Etcheverry clan, one of the first families in Uruguayan wine, arrived from the Basque village of Hasparrieu, a part of the region known more for grain than for grapes. On the recommendation of a cousin in Cadillac, France, who had a vineyard, they decided to try growing grapes in Uruguay, and so they planted in Las Piedras and began a multigenerational business. In 1963, they ventured into Concordia, near Rodríguez in neighboring San José, and took over a decrepit vineyard considered among the oldest in the country. Today wine is made and sold by the third and fourth generations of the family. The wines go under the name of CataMayor,

Basque for "good taste," and not Castillo Viejo, as the earliest wines made under the eponymous label were inexpensive and of lower quality.

Bouza

Founded: 1942

Address: Camina de la Redención 7658 bis, Montevideo, www.bodegabouza.com

Owner: Juan Francis Bouza

Winemaker: Eduardo Boido

Known for: Tannat

Signature wine: Tannat Reserva

Other labels: Parcela Única, Monte Vide Eu

Visiting: Open to the public; tasting room; restaurant

Uruguay's most critically acclaimed producer is a small winery located close to Montevideo in Melilla, with additional plantings in Las Violetas. The winery property dates back to 1942, but it was acquired and restored by Juan Bouza in 2002 with money obtained from selling his food business. Among other culinary claims to fame, Bouza was the largest purveyor of frozen empanadas in all of South America. Eduardo Boido, a partner in the winery, is widely considered one of the country's most gifted winemakers. He was Uruguay's first doctor of enology and teaches at the University of the Republic. Eduardo's wife, Katarina, is also a talented winemaker and consults for two wineries in Argentina. If the quality of the Bouza wines isn't enough of a lure, the Bouza property houses a top-notch restaurant (complete with an excellent sommelier) and a world-class collection of antique cars and motorcycles housed in a small, tasteful museum. A small inn is planned in the near future.

Bracco Bosca

Founded: 2005

Address: Carretera Sosa Díaz, km 43.500, Piedra del Toro, Canelones; no website

Owners: Bracco family, Mirtha Bosca

Winemakers: Edgar Barrera, Joaquín Bosca

Viticulturist: Enrique Sartore

Known for: Tannat, Petit Verdot, Syrah

Signature wines: Bracco Bosca Tannat, Cabernet Sauvignon

Visiting: Open to the public by appointment only; tasting room

Located in Atlántida and with four generations of family history, this winery has a distinctly Piemontese feel, which is not surprising given that the Bracco side of the clan comes from Spigno Monferrato and the Bosca side from Trezzo Tinella. Until recently they were exclusively growers, selling their fruit to local wineries, but they decided to try making their own wine. Undoubtedly the late founder's vivacious and savvy daughter, Fabiana Bracco, export director of Finca Narbona, had a say in that decision. She assists with some winery operations. With thirty acres of vineyards planted on clay and limestone soils, the winery has an unexpected and intriguing focus on Petit Verdot and Syrah.

De Lucca

Founded: 1940

Address: Ruta 48, km 13, El Colorado, Canelones, www.deluccawines.com

Owner: Reinaldo De Lucca

Winemaker: Reinaldo De Lucca

Consulting winemaker: Sylvaine Leplatre

Viticulturist: Reinaldo De Lucca

Known for: Tannat, Merlot, Syrah, Nero d'Avola, Cabernet Sauvignon, Marsanne, Sauvignon Blanc

Signature wine: Río Colorado

Visiting: Open to the public by appointment only; tasting room

The De Lucca winery and its proprietor, Reinaldo De Lucca, are important to Uruguay. As a leading proponent of El Colorado, a *canario* terroir celebrated for its limestone soils and its location in the highest part of Canelones, Reinaldo De Lucca is, according to Julia Harding, "probably the most highly trained and internationally experienced viticulturist in Uruguay." The family is half Tuscan and half Piemontese. Reinaldo's father started producing wines in 1940: in a convenient arrangement, he ended up marrying the daughter of the grower from whom he bought grapes. Reinaldo can easily be considered as important in Uruguayan viticulture as Pedro Parra is in Chile. He's an awfully good winemaker, too.

Familia Irurtia

Founded: 1913

Address: Ruta R. 97, km 2.300, Cerro Carmelo, Colonia, www.irurtia.com.uy

Owner: Irurtia family

Winemaker: Carlos Irurtia

Viticulturist: Marcelo Irurtia

Known for: Tannat

Signature wine: Dante Irurtia de Colección

Other labels: Viñagala, Km. 0, Río de la Plata Reserva

Visiting: Open to the public by appointment only; tasting room

Founded by Basque immigrants who emigrated from Uitzi in the mid-1800s, Irurtia is a true family affair, with third-generation Dante Irurtia and his three sons running the show. The family were originally stonemasons and settled in the area near Serra de Carmelo to work in the quarries. Although they had no winemaking experience, they used their earnings to buy vineyard land. Today they own eight hundred acres of vineyards, constituting one of the larger landholdings in Uruguay. Since their first vintage in 1913, they have continued to expand and modernize. Irurtia is acknowledged for both wines and grappa.

Filgueira

Founded: 1999

Address: Ruta 81, km 7, Cuchilla Verde, Canelones, www.bodegafilgueira.com

Owner: Necchini family

Winemaker: Melissa Barrera

Viticulturist: Gustavo Díaz

Known for: Tannat, Sauvignon Gris

Signature wine: Famiglia Necchini

Other labels: Patio Sur, Fuga

Visiting: Open to the public by appointment only; tasting room

Located in the Santa Lucía area of Canelones and with roots that go back to Galicia in the early twentieth century, the Filgueira winery is as close as you get in Uruguay to the French château concept, with the winery vineyards completely surrounding the facility. The current owners have had the winery since 2010, though it was established a dozen years earlier. Their Sauvignon Gris, a rarity in South America, came in accidentally with other vines imported from France and has become something of a cult wine.

Finca Narbona

Founded: 1998

Address: Ruta 21, km 268, Carmelo,
   Colonia, www.narbona.com.uy

Owner: Eduardo Cantón

Winemaker: Valeria Chiola

Consulting winemaker: Michel Rolland

Viticulturist: Valeria Chiola

Known for: Tannat, Pinot Noir

Signature wine: Luz de Luna Tannat

Other labels: Luz de Luna

Visiting: Open to the public by appointment only; tasting room; restaurant; inn

In the vanguard of Uruguayan wineries, Finca Narbona is the baby of Eduardo Cantón, a wealthy and visionary Argentinean, who purchased a vast amount of land around Carmelo in 1990. He has been gradually developing it over the past dozen years: it now features a gated real-estate community with sublime homes, a Four Seasons hotel, and his pride and joy, the historic winery with no expense spared in its restoration. The winery building dates back to 1740, making it one of the oldest buildings in Uruguay. The *estancia,* which was established by Juan de Narbona in 1909, includes a charming restaurant, vineyard, farmhouse, and lodge about ten minutes from the Four Seasons. Narbona is part wine sanctuary and part foodie heaven: in addition to its fine wines, it produces its own honey, olive oil, cheese, and numerous other offerings that are served in the restaurant.

Giménez Méndez

Founded: 1940

Address: José Batlle y Ordóñez 165,
   Canelones, www.gimenezmendez.com

Owner: Marta Méndez Parodi

Winemakers: Mauro Giménez Méndez,
   Luis Giménez Méndez

Viticulturist: Jorge Fernández

Known for: Tannat, Malbec, Syrah,
   Cabernet Sauvignon, Sauvignon Blanc

Signature wine: Luis A. Giménez Super
   Premium Tannat

Other labels: Las Brujas, 100 Años

Visiting: Open to the public by appointment only

This winery does an admirable job of seamlessly connecting a traditional Uruguayan palate with a modern taste sensibility. Marta Méndez Parodi, a former president of Wines of Uruguay, is so beloved that she was asked to stay on past the end of her term, although she declined. Taking over from her late husband, she has guided the winery with a steady hand and, along with her two sons, managed to retain its position as one of the most popular wineries with the local population. This is due in part to its location right in the center of the town of Canelones and in part to the fact that it still makes ample amounts of common wine to complement the *vinifera* offerings. Their Puzzle wine is unique: a blend of fifteen grape varieties, eleven red and four white, from different vineyards.

Grupo Traversa

Founded: 1956

Address: Av. Pedro de Mendoza 7966, Montevideo, www.grupotraversa.com.uy

Owner: Traversa family

Winemaker: Alejandro Gatto

Viticulturist: Fernando Traversa

Known for: Tannat, Merlot, Cabernet Sauvignon, Cabernet Franc, Syrah, Chardonnay, Sauvignon Blanc

Signature wine: Viña Salort

Other labels: Asolo, Faisán

Visiting: Open to the public by appointment only

One of Uruguay's larger wineries, Grupo Traversa is young but with old roots (the family came over in 1904 from Italy). Initially the family also grew strawberries and Muscat grapes. Today they are committed to a range of *vinifera* varieties, along with a large quantity of common wines for the day-to-day market: they have about a third of Uruguay's market for common wines. Grupo Traversa is developing a focus on exports, and specifically the Asian market. It was the first Uruguayan winery to participate in Vin Expo Hong Kong in 2008 and to establish a presence in Japan.

H. Stagnari

Founded: 2000

Address: Ruta 5, km 20, La Paz, Canelones, www.stagnari.com.uy

Owners: Héctor Stagnari, Virginia Moreira

Winemaker: Héctor Stagnari

Viticulturist: Héctor Stagnari

Known for: Tannat

Signature wine: Tannat Viejo

Other labels: Selección La Puebla, La Caballada, Dinastía

Visiting: Open to the public by appointment only; tasting room; restaurant by reservation only

Like the Mondavis in California, the Stagnari family is prominent in Uruguay's wine scene. Héctor Stagnari left his family's operation (Antigua Bodega Stagnari) in 2001 to establish his own winery, at which point the family added "Antigua" to the name of the original property to distinguish between the two operations. The winery purchases red fruit in Salto to complement the grapes from Canelones. Héctor Stagnari spent time in both Châteauneuf-du-Pape and Bordeaux before launching his estate.

Juanicó (Establecimiento Juanicó— Familia Deicas)

Founded: 1830

Address: Ruta 5, km 39.200, Juanicó, Canelones, juanico.com

Owner: Deicas family

Winemakers: Adriana Gutiérrez, Santiago Deicas

Consulting winemakers: Paul Hobbs, Peter Bright, Duncan Killiner

Viticulturists: Luis Pua, Gustavo Blumetto

Known for: Tannat

Signature wine: Preludio Barrel Select

Other labels: Don Pascual, Bodegones del Sur, Familia Deicas, Casa Magrez, Pueblo del Sol (U.S.A.)

Visiting: Open to the public; tasting room; restaurant by reservation only

Juanicó, named after the town where it is located, was once owned and run by the state and for many years enjoyed the privilege of being allowed to bottle locally produced cognac, using the name under license from the French. Today the Deicas family, who bought the winery in 1979, has taken this operation to new heights. The winery has many Uruguayan "firsts" under its belt: the first iconic wine ("Preludio," launched in 1992), the first wine exports (a full container to London in 1994), the first international gold medal (at a Spanish wine competition in 1996), the first new wine category (*roble*, for oak-aged wines of quality, in 1996), and the first ISO organic certification in Uruguay (in 1998). Juanicó's collaboration with France's Bernard Magrez, Gran Casa Magrez, has had good European sales and was also a first (as an international joint venture project), with the wines initially released in 1999.

Leonardo Falcone

Founded: 1886
Address: Av. Wilson Ferreira
    Aldunate y Young, Paysandú,
    www.bodegaleonardofalcone.com.uy
Owner: Falcone family
Winemakers: Carolina Falcone, Cecilia
    Falcone, Sebastián Misuraca
Known for: Tannat, Merlot, Cabernet
    Sauvignon, Cabernet Franc, Chardonnay, Isabela
Signature wine: Leonardo Falcone 120 Años
Other labels: Abuelo Domingo, Santa
    Cecilia

Visiting: Open to the public by appointment only; tasting room

A family operation whose founders emigrated from Basilicata in Italy in the late 1800s, the Falcone winery is located on the land granted to the family by the government when they arrived in Paysandú. Falcone is the lone producer of *méthode ancestrale* sparkling wines that I know of in Uruguay. Using a process by which the carbon dioxide from the original fermentation is captured in the bottle, the *méthode ancestrale* is associated primarily with southwest France and Gaillac. Alas, the wine is available only at the winery.

Los Cerros de San Juan

Founded: 1854
Address: Ruta 21, km 213.200, Colonia,
    www.loscerrosdesanjuan.com.uy
Owner: Los Cerros de San Juan SA
Winemaker: Ana Zapata
Consulting winemaker: Rodolfo Bartora
Viticulturist: Gonzalo Chavarría
Known for: Tannat, Cabernet Sauvignon,
    Merlot, Tempranillo, Pinot Noir,
    Riesling, Sauvignon Blanc, Chardonnay,
    Gewürztraminer
Signature wine: Cuna de Piedra
Other labels: San Juan Fiesta, Maderos
    de San Juan, Jubileum, Lahusen 1854,
    Soleado
Visiting: Open to the public by appointment only; tasting room; restaurant

One of the original and more traditional Uruguayan wineries, Los Cerros de San Juan recently decided to make a quality push. To do so, it hired the well-respected

Álvaro Lorenzo of Alto de la Ballena as a general consultant. Los Cerros de San Juan is well established and was one of the market leaders in the mid-1900s. The winery, south of Carmelo, is in an old stone building with a stone cellar cut directly into the adjacent hillside.

Los Nadies (Bodega Almacén)

Founded: 2011

Address: Ruta 81, km 7, Santa Lucía, Canelones, www.bodegalosnadies.com

Owner: Manuel Filgueira

Winemaker: Manuel Filgueira

Viticulturist: Manuel Filgueira

Known for: Tannat, Merlot

Signature wine: Equilibrio

Other labels: Ímpetu Picardía

Visiting: Open to the public by appointment only; tasting room; restaurant

An upstart operation whose name means "the nobodies" in Spanish, Los Nadies is irreverent in approach and was born almost spontaneously when Manuel Filgueira and a few friends took a chance on purchasing some Tannat and Merlot grapes and making a go of it. The philosophy of the "group of friends," as they prefer to be identified, is one of minimal hierarchy and an everyone-does-everything approach. As Filgueira's family owned an eponymous winery, you could say it's in Manuel's blood.

Marichal

Founded: 1938

Address: Ruta 5, km 39, Echevarría, Canelones, www.marichalwines.com

Owners: Juan Carlos Marichal, Lidia Santos de Marichal

Winemakers: Sebastián Strada, Alejandro Marichal, Juan Andrés Marichal

Viticulturists: Juan Carlos Marichal, Alejandro Marichal

Known for: Tannat

Signature wine: Marichal

Visiting: Open to the public by appointment only; tasting room

Dating back to 1916, the Marichal family, now in its fourth generation, has its roots in the Canary Islands and Italy. They originally planted and grew grapes, apples, and peaches, selling the fruit to the Giménez Méndez operation, until they decided to start their own winery in the late 1930s. Located in the subregion of Echevarría, a small subsection of Las Violetas in Canelones, Bodega Marichal is innovative and focuses on wines that other local producers do not, including Pinot Noir and sparkling wine. Over 50 percent of their production is exported, mostly to Brazil.

Pagos de Atlántida

Founded: 2011

Address: Ruta 11, km 153, Canelones, pagosdeatlantida.com

Owner: Pedro González

Winemaker: Marcos Carámbula

Viticulturist: Marcos Carámbula

Known for: Tannat, Viognier, Verdello

Signature wine: Pagos de Atlántida Tannat

Other labels: Tenique

Visiting: Open to the public by appointment only; tasting room

Atlántida, midway between the cities of Montevideo and Maldonado and at the eastern edge of the Canelones region, feels and looks more like the Maldonado region. One of just a handful of wineries to focus on Verdello in addition to Tannat (which is almost de rigueur in Uruguay), the *garagiste* bodega is part of a holistic agricultural operation that is as proud of its livestock (Corriedale sheep) and other native crops as it is of its wines. Its mountain cherries and guavas are also reported to be outstanding.

Pisano

Founded: 1924

Address: Ruta 68, km 29, Progreso, Canelones, www.pisanowines.com

Owners: Daniel Pisano, Eduardo Pisano, Gustavo Pisano

Winemaker: Gustavo Pisano

Viticulturist: Eduardo Pisano

Known for: Tannat, Petit Verdot, Syrah, Pinot Noir, Cabernet Franc, Torrontés, Viognier

Signature wines: ArretXea, Axis Mundi

Other labels: 5th Generation, Río de los Pájaros, Cisplatino

Visiting: Open to the public by appointment only; tasting room

Immigrants from Liguria in the late 1800s, the Pisanos are considered a national first family of wine. The winery is now run by three brothers. Daniel Pisano has taken the winery global, while Gus-

tavo ensures that the high quality is in the bottle and Eduardo that the fruit is pristine. Despite its modest production volume, Pisano is exported to more countries than any other Uruguayan wine.

Pizzorno

Founded: 1910

Address: Ruta 32, km 23, Canelón Chico, Canelones, www.pizzornowines.com

Owner: Carlos Pizzorno

Winemakers: Carlos Pizzorno, Marcelo Laitano

Consulting winemaker: Duncan Killiner

Viticulturist: Carlos Pizzorno

Known for: Tannat, Sauvignon Blanc

Signature wine: Primo

Other labels: Don Próspero

Visiting: Open to the public by appointment only; tasting room

With Piemontese roots, the Pizzorno family consistently offers some of Uruguay's best wines. Now in the third generation, the family operates a boutique-style winery in Canelón Chico, with Carlos and his wife, Ana, involved in every detail. When Ana is not managing exports and sales, she works part-time as a doctor specializing in colorectal cancer, with a research interest in the beneficial effects of wine consumption. Their son, Francisco, is learning the trade by working on overseas vintages. The single vineyard is divided into a dozen parcels, each of which is vinified and managed separately—an unusual approach in Uruguay.

Santa Rosa

Founded: 1898

Address: Cesar Mayo Gutiérrez 2211, Montevideo, Canelones, bodegasantarosa.com.uy

Owners: Passadore and Mutio families

Winemaker: Juan Pablo Fitipaldo

Consulting winemaker: Jean Pierre Lede

Viticulturist: Andrés Passadore

Known for: Tannat

Signature wine: Tannat del Museo

Other labels: Del Museo, Juan Bautista Passadore, Punta del Este, Chateau Tierry, Medio & Medio

Visiting: Open to the public; tasting room

Five generations of Passadores and Mutios have been involved in the wine business in Uruguay. The Passadores originally settled in Colón, a small city located a few miles north of Montevideo, and began producing common wine under the brand name Santa Rosa. As business took off, they hired Ángel Mutio, an accountant and the son of Basque immigrants. Soon afterward, Yolanda Passadore married Ángel, linking the destinies of the two families. They produced Uruguay's first sparkling wine in 1936, using the *méthode traditionnelle,* and named it Fond de Cave. It became a tradition and remains one of the best-selling domestic sparkling wines in the Uruguayan market.

Toscanini (Juan Toscanini e Hijos)

Founded: 1908

Address: Ruta 69, km 30.500, Canelón Chico, Canelones, toscaniniwines.com

Owners: Nelson Toscanini, Pedro Toscanini, Jorge Toscanini

Winemakers: Nelson Toscanini, Pilar Sotelo, Juan Pablo Toscanini

Viticulturist: Pedro Toscanini

Known for: Tannat, Cabernet Sauvignon, Merlot, Cabernet Franc, Sauvignon Blanc, Chardonnay, Viognier

Signature wine: Adagio Expressivo

Other labels: Alma Joven, Rendibú

Visiting: Open to the public by appointment only; tasting room

The roots of the Toscanini winery, located in Canelón Chico, go back to Genoa in 1894, when Juan Toscanini and his wife, María Bianchi, immigrated to Uruguay. Now in its fourth generation, the family business is based on grapes obtained from 185 acres of vineyards in Canelón Chico and nearby Paso Cuello. In addition to a full range of table wines, they make three unique wines, two of which are dessert wines under the Rendibú label. Especially distinctive is the purposely oxidized "Vino de Solera," a blend of Chardonnay and Sauvignon Blanc exposed to the sunlight in cask, which gives it a Madeira-like taste.

Viña Progreso

Founded: 2008

Address: Ruta 68, km 29, Progreso, Canelones, www.vinaprogreso.com

Owner: Gabriel Pisano

Winemaker: Gabriel Pisano

Viticulturist: Several working under long-term contracts

Known for: Tannat, Sangiovese, Cabernet Franc, Pinot Noir, Syrah, Viognier

Signature wine: Sueños de Elisa

Visiting: Open to the public by appointment only; tasting room

The apple does not fall far from the tree. Gabriel Pisano is the nephew of Daniel Pisano of the eponymous winery. A true millennial winemaker and one who embraces the innovative, Gabriel, based in Progreso, has focused on unique blends, different approaches, and a playful style of packaging and labeling that would be familiar to many in the United States but is considered a bit radical in Uruguay. In addition to handling his wines, he still works part-time as a winemaker in the family winery. And true to the Pisano tradition, he is already successfully exporting his own wines.

Viña Varela Zarranz

Founded: 1888 (vineyard); 1944 (winery)

Address: Ruta 74, km 29, Suárez, Canelones, www.varelazarranz.com

Owner: Varela family

Winemaker: Enrique Varela

Viticulturist: Ricardo Varela

Known for: Tannat, Muscat

Signature wine: Tannat Crianza

Other labels: Roble, María Zarranz, Topacio

Visiting: Open to the public by appointment only; tasting room

The current Varela Zarranz winery sits on the same spot in Suárez as it did in 1944, when some sixty acres were purchased from Diego Pons, a pioneer vintner along the lines of Harriague and Vidiella, and a winery was established. The estate now covers 240 acres, comprising the winery, a variety of vineyards, and a spectacular botanical garden that was the dream of the Varela family. The brothers Ricardo and Enrique are in charge of grapes and wines, respectively, and Ricardo's daughter Victoria oversees the exports and serves as winery ambassador. In addition to a line of sparkling wines, Varela Zarranz produces the country's only varietal Muscat à Petits Grains, vinified as a dry wine. The operation works exclusively with French varieties.

Viñedo de los Vientos

Founded: 1998

Address: Ruta 11, km 162, Atlántida, Canelones, www.vinedodelosvientos.com

Owners: Pablo Fallabrino, Mariana Cerutti

Winemaker: Pablo Fallabrino

Viticulturist: Pablo Fallabrino

Known for: Tannat, Gewürztraminer, Chardonnay, Nebbiolo

Signature wines: Alcyone Tannat, Angel's Cuvée Ripasso de Tannat

Other labels: Eolo, Estival, Catarsis

Visiting: Open to the public by appointment only; tasting room; restaurant by reservation only

Viñedo de los Vientos is a small, iconoclastic winery with gigantic ideas. The owner, Pablo Fallabrino, inherited the property from his father when he was just twenty-one. The enfant terrible of the Uruguayan wine business, he has surfer looks and a hang-loose attitude and insists on approaching everything differently. A case in point is his "Angel's Cuvée Ripasso de Tannat," made using the traditional Italian method of sun-drying grapes for one month and using the resulting raisins to referment a young wine. The winery boasts thirty-seven acres of Cabernet Sauvignon, Trebbiano, Tannat, Gewürztraminer, Chardonnay, Nebbiolo, and other varieties. Since the first vintage, Pablo Fallabrino has set his sights on the United States market and now sells 90 percent of his wine there. His wife, Mariana Cerutti, works with him and is, among her myriad talents, an excellent chef.

7

BOLIVIA, COLOMBIA, ECUADOR, PARAGUAY, PERU, AND VENEZUELA

Stainless steel vinification vats at the Santiago Queirolo Winery in the village of Pachacamac, south of Lima, Peru. Photo by Mario G. Vingerhoets Pflucker.

Though none of the other wine-producing nations of South America can compete with the leading four, they complete the mosaic of the continent's wine scene and add texture to its offerings. I present them in alphabetical order here.

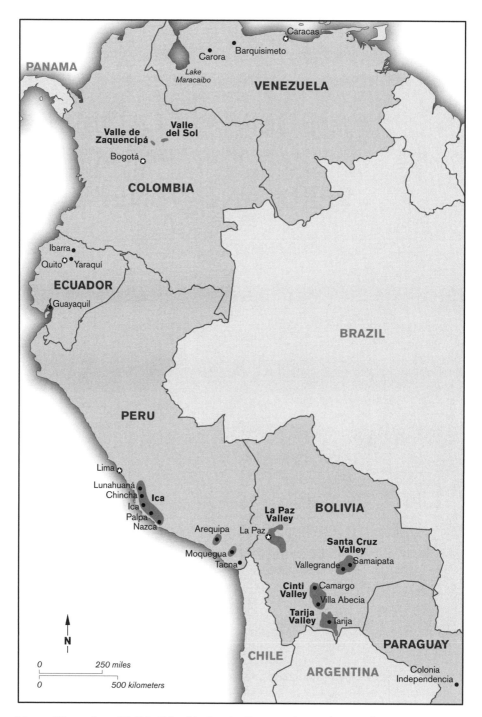

Map 10. Wine regions of Bolivia, Colombia, Ecuador, Paraguay, Peru, and Venezuela

BOLIVIA

Planted acreage: 7,400

World rank in acreage: 63

Number of wineries: 77

Per capita consumption (liters)/world ranking: 2.0/94

Leading white grapes: Chardonnay, Muscat, Semillon

Leading red grapes: Cabernet Sauvignon, Carignan, Malbec, Merlot, Syrah

Memorable recent vintages: 2004, 2006, 2010, 2011

INTRODUCTION

Bolivia is not exactly world renowned for its wine; the agricultural product for which it is best known, the coca leaf, is a topic for other books. Grapes do play a significant role in the country's economy and culture, though more for *singani*, Bolivia's equivalent of pisco, than for wine. Bolivian vines grow at higher altitudes than anywhere else in the world: an average of 8,900 feet (twice as high as the upper reaches of Argentina's Uco Valley), with the fruit zone being between 5,600 and 9,200 feet. Wine-country visitors should take measures to prevent altitude sickness.

Although the vineyards in Bolivia are located between 17° and 22° south, well north of the classic viticultural zones of the Southern Hemisphere, the climate is still temperate and semiarid as a result of the high altitude. Bolivia has numerous small valleys where grapevines have been growing for centuries.

LOCAL HISTORY

Named after the independence fighter Simón Bolívar, Bolivia broke away from Spanish rule in 1825. As in the rest of South America, grapes, notably Muscat of Alexandria and Negra Criolla (also known as Pais or Mission), were introduced to Bolivia by the Spanish conquistadores in the valleys of La Paz (in 1550), Mizque (1580), Camargo (1590), Santa Cruz (1590), and Tarija (1600). The discovery of the silver mines of Potosí, the world's richest, in the 1600s lured treasure hunters from around the globe. By 1630, Potosí's population was greater than that of Paris or London. When local and imported wine was distilled for the miners to protect them from the cold weather, *singani* was born. The name is likely derived from the Bolivian town of Sinkani.

By the early 1900s, French technology was introduced to wineries around Camargo to improve the quality of *singani*. Companies expanded so much that viticultural-sector levies became the primary source of Bolivia's industrial tax income. In the late 1960s, Tarija's Kohlberg winery was the first to invest in Argentinean wine technology for production, marking a tangible step forward in wine quality. Neighboring wineries followed suit, and soon a well-developed table-wine industry was established in Tarija. The cultivation of *vinifera* grapes for premium wines was expanded in 1978 when virus-free vines were brought in from California by the La Concepción winery, which, after twelve years of planting and experimentation, launched its first wines and set the standard for the modern Bolivian wine industry.

BOLIVIAN GAME CHANGERS

JULIO KOHLBERG CHAVARRÍA

The founder of Kohlberg Winery, Julio Kohlberg Chavarría was the first to introduce modern technology into Tarija in the late 1960s, demonstrating that making quality wines was plausible in Bolivia. He was also a pioneer in importing *vinifera* vines and bringing in various selections from Argentina, and in the early 1970s he was at the forefront of investments in new production equipment. He passed away in June 2012 at the age of ninety-one.

SERGIO PRUDENCIO NAVARRO

Winemaker for La Concepción winery from 1990 until early 2014, Sergio has been a front-runner in producing quality varietal, reserve, and super-premium wines. He introduced the use of genetically pure and virus-free vines, imported superior selections and clones of *vinifera* grapes, and accessed a broader selection of rootstocks, better suited to local terroirs, than those previously available. He was also the first to install high-quality, stainless-steel winemaking equipment. In 1994 he launched the unofficial designation *vino de altura* (altitude wine), a concept that recognizes the unique attributes of Bolivia's steep, high-altitude vineyards. It is claimed that these higher locations produce a greater concentration of resveratrol, an antioxidant found in the skins of the grapes. Other Bolivian high-altitude crops, such as coffee and quinoa, are also acknowledged for their superiority. Sergio's recent departure was prompted by his passion for the Bolivian wine industry: he has left to manage a new association of wineries called Wines of Bolivia to promote exports of Bolivian wines and *singani*.

REGIONS AND WINERIES

The planted acreage in Bolivia is 2 percent of that of Argentina; only four of the country's nine departments have vineyards, and of their output, only half is intended for table wine, with the balance used to make *singani*. Of the premium wines produced, 77 percent are red, 20 percent white, and the remaining 3 percent rosé, dessert, and fortified bottlings.

Since 1977, CENAVIT (the National Viticultural Center) has operated in Tarija. It has experimental vineyards, laboratories, and a fully equipped experimental winery. While there is a DO (denomination of origin) for *singani,* there are no legal appellations for wine, though two are pending in Camargo and Tarija.

Like Europeans, Bolivians distinguish wines by region and style rather than by grape variety. The wines have a distinctive style and flavor profile that is likely due to the high altitudes of the vineyards. The greater ultraviolet exposure in the thin air can create an almost roasted character in the fruit. The grapes respond to this exposure by developing much thicker skins and producing phenolic components in their pulp. These produce a distinctive taste, especially as the wines age. As most Bolivian wines are enjoyed young, however, most consumers probably don't notice.

TARIJA VALLEY

The central valley of Tarija is located in the eponymous department in southern Bolivia, on the border with Argentina. Most of the vines are located in the areas of Aviles and Cercado, with small parcels found in Arce and Mendez. Irrigation water comes from the Camacho, Guadalquivir, and Santa Ana rivers in addition to the San Jacinto Dam, which is fed by the Tolomosa River. Beyond the oases of the vineyards, the land is desolate. Spring frosts hang on until mid-September, and the fall frosts arrive in May. As in Argentina's higher-altitude vineyards, hail damage can destroy 8 to 10 percent of Bolivia's annual grape production.

The most widely planted variety in Tarija (and Bolivia) is Muscat of Alexandria, used for table grapes, *singani,* and wine. The most important varieties planted for quality wine production are Cabernet Sauvignon, Syrah, Malbec, Merlot, Tannat, and Barbera for reds and Sauvignon Blanc, Chardonnay, Riesling, and Torrontés for whites. As in the rest of Bolivia, 100 percent of Tarija's vineyards are hand harvested.

RECOMMENDED PRODUCERS

Aranjuez	Kuhlmann
Campos de Solana	La Concepción
Casa Grande	Magnus
Kohlberg	Sausini

CINTI VALLEY

The Cinti Valley encompasses the area around the town of Camargo in the north and a smaller area around Villa Abecia in the south. The viticulture of Camargo has strong ties to historical cultivation practices. Most vineyards are still planted with criolla varieties brought by the Spaniards, notably Muscat of Alexandria and Negra Criolla. The vines grow around *molle* (pepper) trees that, together with the beautiful landscape and its viticultural history, led UNESCO to consider classifying the area as a viticultural heritage site. Geographically, Camargo is located just off the national highway connecting Tarija to Potosí, the world's highest city. Vineyards here are planted at altitudes between 7,700 and 9,200 feet. Today there are only 750 acres under vine (though estimates suggest that 15,000 acres have the potential for grape cultivation), but with an increasing focus on the production of quality wines.

RECOMMENDED PRODUCERS

La Casona de Molina

Viñedo Los Infinitos

SANTA CRUZ

Bolivia's largest department, Santa Cruz is located southwest of the Cordillera de los Andes between 18° and 19° south. While there are just a couple of wineries here currently, most locals anticipate larger investments in Santa Cruz in the not-so-distant future. The two key areas here are Vallegrande and Samaipata. The latter, whose name means "the height of rest" in the Quechua language, is a favorite tourist spot with the locals.

RECOMMENDED PRODUCER

Uvairenda

COLOMBIA

Planted acreage: 6,400

World rank in acreage: 67

Number of wineries: 4

Per capita consumption (liters)/world ranking: 1.12/152

Leading white grapes: Chardonnay, Muscat, Riesling Italico

Leading red grapes: Cabernet Sauvignon, Merlot, Pinot Noir (mostly for sparkling wine)

Memorable recent vintages: n/a

INTRODUCTION

My initial reconnaissance suggested that Colombia had very little to offer in the way of wine. When I consulted my South American peers, they all shrugged their shoulders. But my project manager, Connie, spoke to her cousin, who had been to Cartagena and had been pleasantly surprised by the food and wine culture there. She did some digging, and we soon found out about the popular Bogotá food and wine festival (www.bogota wineandfood.com) and Colombia's VinoExpo (http://expovinos.exito.com). I was sold.

Although Colombian consumption of wine remains low, at just over one liter annually per capita, it has nevertheless doubled since 2006 and is formidable when compared to the statistics for neighboring Ecuador, Venezuela, and Peru. Apparently younger people, in particular, are developing a taste for wine, encouraged by wine festivals and in-store tastings, though beer and rum remain the beverages of choice.

Straddling the equator, Colombia has a climate that is not conducive to wine production, and volumes are not significant. The wine lists of restaurants in Cartagena or Bogotá consist mainly of offerings from Chile, Argentina, and Spain. Imports, and not domestic wines, account for the increase in wine consumption.

COLOMBIAN GAME CHANGERS

MARCO QUIJANO

A globally acknowledged expert on high-altitude tropical viticulture and enology, Marco Quijano founded Marqués de Puntalarga, a winery specializing in German varieties, where he has cultivated quality fruit in the tropical conditions of the Valle del Sol since the early 1980s.

PABLO TORO

Toro, the founder of Marqués de Villa de Leyva, is celebrated for pioneering viticulture in the high-altitude area of Sutamarchán (in the department of Boyacá, a part of Ricaurte Province). He has been at the forefront of establishing and improving vineyard practices for the production of *vinifera* grapes in the tropics.

REGIONS AND WINERIES

BOYACÁ

The department of Boyacá has two zones: the Valle del Sol (Valley of the Sun) and the Valle de Zaquencipá, which includes three municipalities: Villa de Leyva, Sutamarchán, and Sáchica. Both regions are defined by mountains that hold the cloud cover, resulting in balanced sun penetration and mitigation of the heat. Bodegas del Rhin, Vinos de la Corte, and Casa Grajales, all domestic companies focused on bulk wine, accounted for 17 percent of national sales by volume in 2011.

RECOMMENDED PRODUCERS

Marqués de Puntalarga
Marqués de Villa de Leyva

ECUADOR

Planted acreage: less than 100 acres

World rank in acreage: n/a

Number of wineries: 3

Per capita consumption (liters)/world ranking: .29/150

Leading white grapes: Chardonnay, Palomino, Viognier

Leading red grapes: Barbera, Cabernet Sauvignon, Pinot Noir

Memorable recent vintages: n/a

INTRODUCTION

The name *Ecuador* seems to say it all, but there's more to this country than its latitude. Its attractions range from the famous Galápagos Islands to the world's highest capital city, Quito, nestled in the Andean peaks at 9,350 feet. And even though wine is not nearly as renowned as those marvelous Galápagos tortoises, it has a presence.

Like Colombia, Ecuador imports quite a bit of wine from Chile and Argentina. However, it also produces its own grapes, grown at around eight thousand feet near Quito and in the coastal province of Guayas. Conditions here are challenging and very wet. Dick Handal, an American who owns the Chaupi Estancia winery, said that a few decades ago, there were two distinct weather patterns—a wet season and a dry one—and winegrowing was easier to manage. Now, he laments, it rains almost every day around Quito and is wetter still in Guayas. The wineries humbly call themselves experimental, but they sell quantities of their wines locally and are not thought of as curiosities in the domestic market.

As in the other equatorial wine regions of Colombia, Venezuela, and northern Brazil, semiannual harvesting is the way of life. The wetter weather makes it challenging, but the Ecuadorians manage to pull it off. According to Dick Handal, to achieve more consistent and earlier bud break in the wet environment, Ecuadorian wineries apply a vineyard additive called Dormex just before the natural bud break.

REGIONS AND WINERIES

PICHINCHA

Pichincha Province is located in the northern sierra region and is home to both the city of Quito and an active volcano from which the province takes its name. The vineyard area is about twenty-five miles from the city, situated less than a mile from the new international airport. This makes it easily accessible to local and international tourism, the primary source of winery sales and sustenance. This location, one thousand feet lower than Quito proper and with volcanic, sandy soils, is ideal for Palomino (best known for its role in the production of sherry in Spain's Andalucía region) and Viognier. In reds, Pinot Noir can be lovely.

RECOMMENDED PRODUCER

Chaupi Estancia

GUAYAS

The most populous and the fourth-largest province in Ecuador, coastal Guayas is home to the country's largest city of Guayaquil. The vineyard area is in the Peninsula de Santa Elena, where there is limestone in the soil. Chardonnay and Cabernet Sauvignon are the most consistently successful grapes so far.

RECOMMENDED PRODUCER

Dos Hemisferios

IMBABURA

Imbabura Province and the city of Ibarra sit in Ecuador's northern desert, a short distance northeast of Quito. There is some grape activity here, and Mauricio Davalos, a former economic policy minister, has grown successful experimental plantings in this drier region, though the *vinifera* plantings have been on a modest scale.

PARAGUAY

Planted acreage: 1,200
World rank in acreage: 76
Number of wineries: 1
Per capita consumption (liters)/world ranking: 5.23/74
Leading white grapes: Riesling, Sylvaner
Leading red grapes: Dornfelder, Grenache
Memorable recent vintages: n/a

INTRODUCTION

A small, landlocked country, encircled by Brazil, Bolivia, and Argentina and with only one winery of consequence, Paraguay ranks with Colombia as the least known of all South American wine countries.

In the 1920s and 1930s, German immigrants came to settle the agricultural interior of the country. Many came from Baden and brought viticulture. They found the hot climate difficult for grape growing, and while local wines are acceptable to the domestic market, they have never been regaled internationally. Although a few wines are considered to be of fine-wine quality and bottled, the prevalent business is in bulk wine, which is rarely consumed by itself but rather is mixed in equal parts with cola and ice to make a *par* or a *pareja* (pair or couple).

The Paraguay River divides the country into two regions, and most of the population is based in the east, around the capital city of Asuncíon. The Gerhard Bühler Winery (Vista Alegre), in eastern Colonia Independencia, is the only show in town. Even Herr Bühler would concede that wine is not a national beverage. Paraguayans drink lots of beer and, like their Brazilian neighbors, ample amounts of maté, the tea made from yerba maté leaves.

PERU

Planted acreage: 27,200

World rank in acreage: 51

Number of wineries: 9

Per capita consumption (liters)/world ranking: 0.47/133

Leading white grapes: Chenin Blanc, Muscat, Sauvignon Blanc

Leading red grapes: Malbec, Merlot, Petit Verdot, Tannat

Memorable recent vintages: n/a

INTRODUCTION AND HISTORY

In 1412, according to Inca tradition, the emperor Pachacutec, the architect of Machu Picchu, expanded the Inca Empire into the Ica Valley. While inspecting the region, he fell in love with a gorgeous young maiden. He proposed marriage, but she declined in favor of sticking with a current boyfriend. Apparently benevolent of heart, Pachacutec wanted to demonstrate his love and offered her the gift of her choosing. Her wish was to witness the waters from the mountain rivers reach her town in the desert. Supposedly it took forty thousand workers to build the aqueducts, *la achirana del Inca,* to transport water to her town and the rest of the Ica Valley. It was Pachacutec's irrigation system that provided the Spanish with the water needed to cultivate their first vineyards and produce their first wines.

By the end of the sixteenth century, Peru ruled South American wine commerce, but those halcyon days are long over. Peru's wine industry has been dwarfed by that of its neighbors. Wine has been upstaged by pisco, and the excitement in the food and wine culture of Peru is mostly culinary, as exemplified by South America's leading celebrity chef, Gastón Acurio of Lima. Pisco production is massive and based mostly south of Lima, especially in Ica, where vineyards produce Quebranta grapes. Though the pisco industry is still smaller than Chile's, there are more than four hundred Peruvian producers, dwarfing efforts in fine wine, and their best offerings are regarded as the benchmark by aficionados.

Much of Peru's economy is associated with its coast. Peru's fishing waters are considered the world's richest, and Peru is the world's largest exporter of fish meal, used in animal feed and organic fertilizers. Most of the country's vineyards lie on the coast, with the table-grape business situated north of Lima in Ancash and Trujillo and the quality *vinifera* fruit being cultivated mostly in the central and southern coastal areas. The best wine-producing vineyards can be found near the town of Ica, an oasis about 185 miles south of Lima. Coastal Peru is a desert, intersected by a series of valleys flowing from the Andes to the sea. Rainfall is nonexistent, and vines need to be irrigated via underground aquifers. An added problem in Peru is that *garúa*, coastal sea mists, permeate the vineyards, drawn from the ocean by the warmer inland reaches, and often create mildew problems. Coupled with warm winters (which mean that vines never go completely dormant), these conditions can make growing quality grapes problematic. However, cool currents of offshore air create a desirable counterbalance to the humidity, and sufficient thermal amplitude accentuates a healthy growing environment. The understanding of such conditions has sparked a new excitement in rekindling Peru's historic production of table wines. But it's an uphill battle, as Peru's wines have generally been of poor quality. With an undiscerning local market and scarce export sales, there has been minimal incentive to invest. Then, in August 2007, a magnitude 8.0 earthquake damaged many wineries around Ica. Some went out of business, but a few of the larger wine producers took the opportunity to rebuild their wineries and invest in state-of-the-art technology, which should help improve their wines in future years. And bringing in top-flight consultants like Argentina's Roberto de la Mota and Chile's Adolfo Hurtado can't hurt.

Pisco rules the roost in Peru. Fine restaurants that would like to feature Peruvian wines are frustrated by the low quality of current offerings compared to imported wines (mostly Argentinean and Chilean) and have instead focused on expanding their range of piscos. This can be mind-bogglingly complex, because the production of pisco has become highly refined, involving a range of choices produced in small batches, single varieties, and so on. Another challenge for wine lovers is the high price of local wines. As my Peruvian friend Soledad Marroquín remarked, "Peruvians buy Peruvian wine, so the wineries charge what the market will bear."

REGIONS AND WINERIES

ICA

The southern region of Ica, 185 miles south of Lima, is flanked by the Andes and miles of towering sand dunes. There are several valleys here, including the Ica Valley, Peru's largest and most important wine-producing area, with several bodegas located not far from the village of Huacachina. The site of the first vineyards in Peru, Ica is an important agricultural region that also excels in the production of cotton, asparagus, and olives. The valley hosts an annual wine festival where you can sample the local offerings of wine and pisco. Grapes are also found in the Chincha and Nazca valleys.

RECOMMENDED PRODUCERS

Ocucaje

Santiago Queirolo

Tabernero

Tacama

Vista Alegre

MOQUEGUA

Moquegua, which means "quiet place" in Quechua, is a small region in southern Peru that extends from the coast to the eastern highlands. According to Prudence Rice in her book *Vintage Moquegua*, the Moquegua River Valley is the name given to the area surrounding the middle of the Osmore River, where grape cultivation has been documented as far back as the late sixteenth century. It was an important source of wine for the Potosí silver miners in Bolivia. There is a "pisco route" that allows the visitor to get to know the old colonial bodegas; this takes in the best vineyards in the region, but although this is an important area for distillates, it is not renowned for fine wine.

VENEZUELA

Planted acreage: 3,400

World rank in acreage: 73

Number of wineries: 1

Per capita consumption (liters)/world ranking: 0.18/144

Leading white grapes: Chenin Blanc, Muscat, Sauvignon Blanc

Leading red grapes: Petit Verdot, Syrah, Tempranillo

Memorable recent vintages: n/a

A trip across South America's wine country would be remiss in not mentioning Venezuela, which many people would assume is simply too close to the equator for growing wine grapes. Again, altitude is the answer, along with thermal amplitude exceeding 30°F and very good soil drainage.

The name *Venezuela* is attributed to the Italian explorer Amerigo Vespucci, who led a 1499 naval expedition along the northwestern coast's Gulf of Venezuela. After reaching the Guajira Peninsula, he noted that several villages (*palafitos*) that the indigenous people had built over the water reminded him of the city of Venice, so he named the region *Venezuola,* "little Venice."

Like those of its equatorial neighbors, Venezuela's vineyards generate two harvests per year. Since the 1990s, the most prominent wine area has been Carora, a town in the state of Lara, fifty-four miles southwest of picturesque Barquisimeto. It is not far from Lake Maracaibo, at the foot of the Sierra de Baragua, a region with a relatively dry tropical climate (receiving twenty-four inches of rain annually). Alas, wine is not popular among Venezuelans, though the better restaurants and stores of Caracas stock a surprisingly broad range of Italian and Spanish offerings, among others. Known for being the most *cerveza*-thirsty people in South America, Venezuelans are beer drinkers first and rum drinkers second. Most of the wine is consumed by expatriates, including the families of Europeans who came to Venezuela after World War II. Immigrants have long been drawn by the country's mineral wealth, which includes oil, aluminum, and iron. Wine labeled as Venezuelan wine is usually made with imported fruit and bulk wine and bottled locally.

RECOMMENDED PRODUCER

Bodegas Pomar

The environs of Hacienda La Compañía at the La Concepción winery in Tarija, Bolivia. Photo by Lilo Methfessel.

BOLIVIA

Aranjuez (Bodegas Milcast Corp.)

Founded: 1976

Address: Av. Ángel Baldivieso No. 1976, Tarija, www.vinosaranjuez.com

Owner: Flia. Castellanos Cortez

Winemaker: Franco Sánchez

Viticulturist: Ernesto Verdún

Known for: Tannat, Cabernet Sauvignon, Merlot, Malbec

Signature wine: Aranjuez Tannat

Visiting: Open to the public by appointment only

Aranjuez is the winery brand of Bodegas Milcast, built in Aranjuez in Tarija. The winery grows fourteen varieties of grapes, of which approximately two-thirds are *vinifera*, in vineyards located outside the town of Santa Ana la Vieja near Aranjuez.

Campos de Solana

Founded: 2000

Address: Carretera a Bermejo, km 9, El Portillo, Tarija, www.camposdesolana.com

Owner: Granier family

Winemaker: Nelson Sfarcich

Viticulturist: José Argumedo

Known for: Cabernet Sauvignon

Signature wine: Colección de Altura

Visiting: Open to the public; tasting room

The result of two families' coming together, Campos de Solana (meaning "sunny fields") was created to produce wine for a successful company that sells perhaps Bolivia's best-known brand of *singani*, Casa Real, which traces its roots back to 1925. The Castellano family of Casa Real brings two decades of winemaking and distillation experience to the operation, and the Graniers are a third-generation family of grape growers and distillers. Campos de Solana is one of the newest and most modern Bolivian wineries.

Casa Grande

Founded: 2006

Address: Calle Sucre 332, Tarija, www.casagrande.com.bo

Owner: Luis Adam Michel Mendoza

Winemaker: Tesfaye Abebe

Viticulturist: Gloria Cartagena

Known for: Cabernet Sauvignon

Signature wines: Reserva Trivarietal, Vino
 Espumante Osadía

Other labels: Altima, Osadía, Burbushhh

Visiting: Open to the public by appointment only; tasting room; restaurant

A newbie on the Bolivian wine scene, Casa Grande is the sole facility in Bolivia with underground caves, used for aging the wines in small barriques. Casa Grande is focused on wines rather than *singani* and cultivates a range of grapes, from Syrah to Chardonnay.

Kohlberg (Bodegas y Viñedos La Cabaña)

Founded: 1963

Address: Av. Jorge Paz Galarza, Barrio San
 Jorge 1, Tarija (no website)

Owner: Kohlberg family

Winemakers: Herbert Kohlberg, Erich
 Kohlberg

Viticulturists: Franz Kohlberg, Helmut
 Kohlberg

Known for: Syrah, Carignan, Grenache,
 Muscat of Alexandria

Signature wines: Vino Fino Tinto de Mesa,
 Gran Reserva Don Julio

Visiting: Open to the public by appointment only

For more than forty years, the Kohlberg family has been a leader in the evolution of Bolivia's fine-wine production. The Kohl-bergs are also locally popular, and the winery is very successful. They source 80 percent of their fruit from their own vine-yards in Santa Ana and have arguably the greatest range of *vinifera* varieties in all of Bolivia. The vineyards are set in a hilly landscape above a valley and have the typ-ical reddish, sandy soils of Tarija. Kohl-berg is one of the few Bolivian winemak-ing operations to reach markets beyond South America.

Kuhlmann

Founded: 1930

Address: Calle Franz Kuhlmann sin
 número, Carretera San Luis, km 4,
 Tarijas, www.bodegaskuhlmann.com

Owner: Kuhlmann & Cia Ltda

Winemaker: Carlos Molina

Viticulturist: Carlos Molina

Known for: Muscat of Alexandria, Chardonnay, Xarel-lo, Tempranillo

Signature wine: Altosama

Visiting: Open to the public by appointment only; tasting room

Kuhlmann, which has a long history of selling *singani,* is now also one of Bolivia's few producers of sparkling wine. Named for its German founder, Franz Kuhlmann, who arrived in Bolivia in 1917, the com-pany is run today by the third-generation Dora Molina and her son Carlos, who is the winemaker; Franz Kuhlmann died in 1977. Although they had their start in Ca-margo, they moved to Tarija as the oper-ation expanded. Kuhlmann's "Tres Estre-llas" *singani* is one of Bolivia's preferred marks. First released in June 2012, their Charmat-method sparkling wines are made from estate-grown grapes.

La Casona de Molina

Founded: 2005

Address: Tarija Road, km 5, Camargo, Nor Cinti, www.lacasonademolina.com

Owner: Mario Molina

Winemaker: Mario Molina

Viticulturist: Ruperto Molina

Known for: Cabernet Sauvignon, Bordeaux-styled reds

Signature wine: Don Martín

Visiting: Open to the public; tasting room; restaurant

A small and well-equipped winery, Casona de la Molina has just a couple of vintages under its belt. The winery buys from vineyards with traditional hundred-year-old vines grown around *molle* trees. Varieties used in the blends include Muscat of Alexandria, Negra Criolla (Pais), Cabernet Sauvignon, Vicchoqueña, and Cereza.

La Concepción

Founded: 1986

Address: 27 km southwest of Tarija, 2 km from Concepción, Valley of Concepción, Tarija, www.bodegaslaconcepcion.com

Owners: Various stockholders of Bodegas y Viñedos de La Concepción SA

Consulting winemaker: Sergio Prudencio Navarro

Viticulturist: Ximena Pinedo

Known for: Cabernet Sauvignon, Syrah

Signature wine: Cepas de Altura Syrah

Visiting: Open to the public; tasting room

A leading winery in Tarija, La Concepción has three vineyard sources: Chagunya

(about 60 kilometers from the town of Tarija), Huayrihuana (about half an hour from the winery), and La Compañia (adjacent to the facility). This last vineyard was planted by missionaries and is said to have some vines that are over two hundred years old. La Concepción also has a 15 percent share of the Bolivian *singani* market.

Magnus

Founded: 2006

Address: Calle Ingavi 244, Torrecillas, Tarija, www.bodegasmagnus.com

Owners: Ernesto Magnus and family

Winemaker: Liz Arancibia de Magnus

Known for: Cabernet Sauvignon, Syrah, Merlot

Signature wine: Magnus

Visiting: Open to the public by appointment only

Although the winery is fairly new, the Magnus family traces its wine roots back to 1928. Still on the original site, the modern boutique winery is run by a husband and wife, Ernesto Magnus and Liz Arancibia, who are passionate about quality and make no more than one thousand cases of any lot. Visitors to the winery in Torrecillas, ten minutes from Tarija city center by car, will find a refreshing contrast to other, larger wine companies in Bolivia. Magnus produces exclusively fine varietal wines in small lots.

Sausini

Founded: 1985 (vineyard), 2001 (winery)

Address: Av. El Temporal sin número, San Luis, Tarija, www.bodegasausini.com

Owner: Hinojosa family

Winemaker: Sebastián Molina

Viticulturist: Bernardino Márquez

Known for: Merlot, Cabernet Sauvignon

Signature wine: Sausini Blend—Reserva de Familia 2008

Visiting: Open to the public by appointment only

A new winery that is developing a very good reputation, Sausini is perhaps the best boutique operation in southern Bolivia. It is equally acknowledged for its *singani* and its fine wines.

Uvairenda

Founded: 2007

Address: Camino a Valle Abajo, next to Hotel El Pueblito, Samaipata, Santa Cruz, www.uvairenda.com

Owners: Luis Peregrín, Ortiz Mercado, Francisco Roig Justiniano

Winemakers: Roberto Ignacio Aguilar Pérez, Francisco Roig Justiniano

Viticulturists: Fidel Sibaute Coronado, Roberto Ignacio Aguilar Pérez

Known for: Cabernet Sauvignon, Syrah, Tannat, Chardonnay, Torrontés Riojano

Signature wine: 1750

Visiting: Open to the public; tasting room

A relative newcomer to fine-wine production, Santa Cruz has historically been known exclusively for older Muscat vines and *singani* production. Uvairenda, whose name is a combination of the Spanish *uva* ("grape") and the Guarani *renda* ("place"), is located in the town of Samaipata, seventy-five miles from the city of Santa Cruz. It is the first boutique estate in a region that has been estimated to have 81,000 acres suitable for the production of fine wine—five times the area of either Tarija or Camargo. To date, Uvairenda has just a few vintages under its belt, so it is too early to rush to judgment. The winemakers have the vision but not yet the wine.

Viñedo Los Infinitos

Founded: 2005

Address: Calle Ossio, Villa Abecia, Sud Cinti; no website

Owner: Cristian Villamor

Winemaker: Cristian Villamor

Viticulturist: Cristian Villamor, Rolando Rengifo

Known for: Cabernet Sauvignon

Signature wine: Camataqui Reserva

Other labels: Monte Azul, Tierra Roja

Visiting: Open to the public by appointment only; tasting room

Viñedo Los Infinitos sits in the south of the Cinti Valley. The winery is quite small and high-tech and sells wines to the best restaurants of La Paz. This area has the potential to produce great wine, but its wineries currently lack vision. Wineries like Los Infinitos, with its Tierra Roja line of wines, demonstrate great promise.

COLOMBIA

Marqués de Puntalarga

Founded: 1982

Address: Autopista Duitama, km 7, Belencito, Nobsa, Boyacá, www.marquesdepuntalarga.com

Owner: Marco Quijano Rico

Winemaker: Sandra Rincón

Consulting winemaker: Marco Quijano
  Rico

Viticulturist: Marco Quijano Rico

Known for: Riesling, Riesling × Silvaner,
  Pinot Noir

Signature wine: Marqués de Puntalarga

Visiting: Open to the public; tasting room

In Boyacá, blistering hot and 8,500 feet
above sea level, it's hard to imagine establishing a viable commercial winery. Marqués de Puntalarga sources fruit from the
Sogamoso Valley. Marco Quijano calls
his winery a "boutique winery" pilot project. He was first inspired to make wine
when traveling to Switzerland and Germany to study nuclear chemistry. When
he returned to Colombia and embarked on
winemaking, he first had to convince local
farmers to stop planting potatoes and cultivate vines, which was not an easy task.
He ultimately planted thirty-seven different grapes before deciding on a smaller
selection, mainly Germanic varieties. The
first vintage was in 1987.

Marqués de Villa de Leyva

Founded: 1988

Address: Vía Villa de Leyva, km 10, Santa
  Sofia, Boyacá, www.marquesvl.com

Owner: Toro Cortés family

Winemaker: Mauricio Camacho

Viticulturist: Segundo Gómez

Known for: Sauvignon Blanc, Cabernet
  Sauvignon

Signature wine: Cabernet Sauvignon
  Reserva Especial

Visiting: Open to the public; tasting room

Marqués de Villa de Leyva is the vision of
Pablo Toro, a Cornell University graduate
who is renowned in Colombia as the godfather of freeze-dried coffee. Toro consid-
ers it his mission to educate Colombians
in fine-wine culture. He founded his vineyard, Ain Karim, in the town of Sutamarchán, as a tribute to the city of Jerusalem.
The name (taken from a historic village in
the hills of Jerusalem) means "spring in
the vineyard." He selected the site, with assistance from a French wine institute, for
its climatic and soil similarities to wine-
growing areas in Chile and Argentina.
After several failed attempts at planting
Cabernet Sauvignon, Sauvignon Blanc,
and Chardonnay, the growers finally realized that the vines had to be planted in
the correct orientation to the sun, and success followed. Under the winemaker Mauricio Camacho, who advocated lower yields
and fewer annual harvests, quality has improved markedly.

ECUADOR

Chaupi Estancia

Founded: 1999

Address: Barrio San Vicente, Yaruquí,
  Pichincha, www.chaupiestancia.com

Owner: Handal family

Winemaker: Héctor Olivares Madrid

Viticulturist: Jorge Durán

Known for: Pinot Noir, Palomino, Mourvèdre, Barbera, Sangiovese

Signature wines: Pinot Noir, Palomino
  Fino

Visiting: Open to the public by appointment only; tasting room

A true labor of love on the part of Dick Handal, Chaupi Estancia was established from experimental vineyards planted in the 1980s, and the winery followed. Having spent many years in the Peace Corps in South America, Dick settled with his family in Ecuador after several other stops and planted thirty-two varieties of grapes on sandy volcanic soils. Over time, the most consistent performers have been Mourvèdre, Barbera, Sangiovese, and Pinot Noir for reds and the very successful Palomino for whites. The winery is especially proud of a cuvée that is made from the best red grapes in a given harvest—always an unpredictable affair given the Ecuadoran climate. Pinot Noir is the only red varietal wine. Though now resident in Sonoma County, California, Dick is in Ecuador two or three times a year.

Dos Hemisferios

Founded: 2004

Address: Av. Juan Tanca Marengo, km 4.5, Guayaquil, Guayas, www.doshemisferios.com

Owner: Guillermo Wright

Winemaker: Abel Furlan

Viticulturist: Hernán Cortegoso

Known for: Cabernet Sauvignon, Malbec, Chardonnay

Signature wine: Paradoja

Tasting: Open to the public by appointment only; tasting room

Located in Puerto San Miguel del Morro in Guayas Province, fifteen minutes inland from the town of General Villamil Playas, Dos Hemisferios and its Argentinean winemaker, Abel Furlan, face great challenges in producing wine at latitude zero and at the edge of the ocean. In addition to the difficult climate, the vineyard's remote location means that the grapes must be transported to a winemaking facility in the city of Guayaquil. All the wines have proprietary names and are red blends ("Paradoja," "Bruma," "Del Morro," and "Travesia"), with the exception of their Chardonnay, labeled "Enigma."

PERU

Ocucaje

Founded: 1898

Address: Pasaje Julio Vega Solis, lot 7, Chorrillos, Lima, ocucaje.com

Owner: Rubini family

Known for: Cabernet Sauvignon, Merlot, Malbec, Chardonnay

Signature wines: Don Virgilio Cabernet Sauvignon Reserve, Colección Rubini Crianza

Visiting: Open to the public; tasting room

Ocucaje is one of the "big three" wineries of Peru, along with Tacama and Santiago Queirolo. Alas, information about it is scarce, as its website is minimal and the winery was not forthcoming when asked for information. In business for over a century and run by eight brothers, the domain makes a range of red wines based on Bordeaux varieties, a single white wine (Chardonnay), and several generic wines, including a sacramental wine. Its main business is in pisco, of which it makes a wide assortment of styles at different quality levels.

Santiago Queirolo

Founded: 1880

Address: Av. San Martín 1062, Pueblo Libre, Lima, www.santiagoqueirolo.com

Owner: Queirolo Targarona family

Winemakers: Jorge Queirolo, Ernesto Jiusan

Viticulturist: Jorge Queirolo

Known for: Sauvignon Blanc, Chardonnay, Malbec, Syrah, Tannat, Cabernet Sauvignon, Petit Verdot

Signature wine: Intipalka

Visiting: Open to the public; tasting room; restaurant (Las Viñas); boutique hotel

Genovese in origin, the Queirolo family facility still resides in the same spot in Pueblo Libre as it did at the end of the nineteenth century, when Santiago Queirolo arrived. An ambitious man, he sold octopus and seafood on a street corner, then began selling pisco and wine from his land in Cañete. His business was so successful that his shop stall became a restaurant, and he expanded his sales of pisco. As time went on, Santiago Queirolo became a brand name for wines and spirits, and the restaurant became a side business. Still one of the dozen best places to eat in Pueblo Libre, the Antigua Taberna Queirolo is the oldest restaurant in the area.

Led by the example of Chilean and Argentinean wineries, Santiago Queirolo made massive investments in its vineyards and winery and was recognized in an industry poll as the number one wine and pisco brand in the country. The name of the wines, *Intipalka*, means "valley of the sun" in Quechua, the language of the Inca Empire.

Tabernero

Founded: 1897

Address: Av. Prolongación Andes Rásuri sin número, Chincha Alta, Ica, www.tabernero.com

Owner: Hermanos Rotondo

Winemaker: Bertrand Jolly

Viticulturist: Carlos Rotondo

Known for: Chenin Blanc, Chardonnay, Sauvignon, Malbec, Cabernet Sauvignon, Merlot, Syrah

Signature wine: Gran Tinto Fina Reserva crianza

Visiting: Open to the public; tasting room

One of a small number of Peruvian wineries with French winemakers, Tabernero has been in the capable hands of the Rotondo family since the 1930s. Alas, after Peru's mid-1960s government expropriation of all agricultural land, it had to be restarted. Peru's largest wine exporter (claiming 85 percent of the export market), the Tabernero facility is currently more celebrated for its award-winning piscos but is making large investments in improving wine quality.

Tacama

Founded: 1889 (vineyards first planted 1540)

Address: Alberto del Campo 449, San Isidro, Lima, www.tacama.com

Owner: Olaechea family

Winemaker: Robert Niederman

Known for: Tannat, Malbec, Petit Verdot, Sauvignon Blanc

Signature wine: Selección Especial

Other labels: Don Manuel, Terroix, Gran Tinto, Blanco de Blancos, Gran Blanco, Albilla D'Ica, Amore de Ica, Sinfonía Quantum, Halcón de la Viña

Visiting: Open to the public; tasting room

Claiming to be South America's first vineyard, Tacama today is solely owned by the Olaechea family, whose French winemaker, Robert Niederman, has been at the helm for every harvest since 1961. At one point the late Émile Peynaud was a consultant for the winery. The pink, fortress-like winery building is located five miles northeast of the town of Ica, and two-thirds of its vineyards are planted to grapes for fine-wine production. It is Peru's sole producer of Tannat.

Vista Alegre

Founded: 1857

Address: Cerro Prieto 274, Urb. San Ignacio de Monterrico, Surco, Ica, www.vistaalegre.com.pe

Owner: Picasso family

Winemaker: Rodolfo Vasconi

Viticulturist: Fernando Farre

Known for: Tempranillo, Malbec, Pinot Blanc

Signature wines: Picasso Tempranillo Crianza, Picasso Malbec Reserva Crianza

Visiting: Open to the public; tasting room

A fourth-generation family operation, Vista Alegre also has an international winemaker—not French but Argentinean. There is a real buzz surrounding the winery's Tempranillo.

VENEZUELA

Bodegas Pomar

Founded: 1985

Address: Carretera Lara-Aulia, km 1, Apartado 33, Carora, Lara, www.bodegaspomar.com.ve

Owner: Lorenzo Mendoza (president, Empresas Polar)

Winemaker: Guillermo Vargas

Viticulturist: Guillermo Vargas

Known for: Sauvignon Blanc, Chenin Blanc, Macabeo, Malvasia, Muscat à Petits Grains, Syrah, Petit Verdot, Tempranillo

Signature wines: Terracota Blanco, Terracota Tinto, Pomar Reserva, Pomar Brut

Visiting: Open to the public by appointment only; tasting room

The only real winery in Venezuela, Pomar is a well-funded joint venture with France's Martell Cognac. The Venezuelan partner, Empresas Polar, is a huge food-and-drinks operation that accounts for 85 percent of the beer market. Pomar is a combination of the two owners' names. Though the winery is not remote, it is a dedicated wine drinker who will make the short drive from the Barquisimento airport, which is itself a forty-minute flight from Caracas.

TOURING SOUTH AMERICAN WINE COUNTRY

Traveling the wine regions of South America is not like wine tourism in other places. In the Napa Valley or New South Wales, all you really have to do is rent a car, add a GPS or a map, and set out. Taking that approach in South America is a ticket to adventure. Many South American wineries are not on the main roads, and they often don't have clear addresses or conspicuous signs. More often than, winery locations are given simply as the name of a highway or road, with or without a street number, or the combination of a road (*ruta*) and kilometer reading (e.g., Ruta 7, km 66.5). The always-challenging "s/n" stands for *sin número* in Spanish and *sem número* in Portuguese, both of which mean no number.

To add to the challenge, many roads are unmarked, and GPS navigation often isn't reliable. And when you do get lost (which you will), your chances of getting directions are slim unless you speak the local language (either Spanish or Portuguese) and are lucky enough to run into someone who knows the local bodegas. But fear not: you can have an enjoyable wine tour in South America if you plan and prepare.

You can do it on your own by obtaining some maps and directions, summoning your reserves of patience and boldness, and heading out onto the wine routes. But do yourself a favor and hire a company to do the driving for you. It's well worth the investment. Below I list English-speaking contacts in the four primary wine-producing countries whom I know personally and who will add nothing but enjoyment to your trip. They know their respective wine areas, are acquainted with many of the producers, and are seasoned pros. And they'll save you from being pulled over by the local police after sampling one wine too many.

> Argentina: Walter Galimberti, Wine Tour Logistics, www.winetourlogistics.com; wgalim@yahoo.com.ar, Ph: (54) 261–6339555; Lencinas 1015, M5539EAT, Las Heras, Mendoza, Argentina

> Brazil: Marcio Dalla Valle, Tchury Translado Executivo Ltda, mdv540@hotmail.com, Ph: (54) 8111–2019; Rua Mateus Valduga, 179, Bairro São Bento, Bento Gonçalves, Rio Grande do Sul, Brazil

> Chile: Alfonso Soto, Alfonso Soto G. Company, alfsot@hotmail.com, Ph: (56) 9 82348907; Loma Alta, 06867, Puente Alto, Santiago, Chile

> Uruguay: Ryan Hamilton, The Wine Experience, www.thewine-experience.com; ryan@thewine-experience.com, Ph: (598) 99675750; Los Cisnes esquina Los Horneros sin número, José Ignacio, Maldonado 20402, Uruguay

For other countries, a travel agent can help you organize a trip. In addition to hiring a driver, I offer the following general recommendations:

- Don't expect to see it all in a week. For most wine visitors, it's a long trip just to reach South America. Once you're there, distances between wine regions are often substantial. It's best to focus on a single region or two rather than attempting to visit several with two days here and three days there.

- Internal flights can be intermittent and unpredictable and are not always the most practical or economical way to go. Consider doing a lot of your adventure by hired car or by bus. Overnight buses aren't just for budget travelers: comfortable versions have fold-out beds, serve you food, host bingo, and even show movies.

- As you head toward the South Pole, the ozone layer gets thinner. The same is true as you ascend to higher altitudes. Sunglasses, sunhats, and sunscreen are essential for avoiding serious sun damage.

- It may be hot in the daytime, but it can be downright chilly at night, especially at higher altitudes. Pack accordingly, dress in layers, and check local weather forecasts before setting out.

- In the Southern Hemisphere, summer and the high season fall around Christmas, so avoid the end of the calendar year at all costs. December holidays are more expensive, and many wineries are closed for their own holidays.

- United States citizens should expect some extra costs. In reciprocation for the hefty visa fees extracted by the United States from their citizens, several of the South American countries impose "reciprocity fees" for American visitors. In Brazil, this fee is charged when you apply for a visa. In Chile and Argentina, which do not require tourist visas, reciprocity fees are paid online (Argentina) or at the airport (Chile) on arrival in the country. Once the fee is paid, the receipt is attached to your passport, and it's good for the life of the passport or ten years. Credit cards and U.S. dollars accepted!

GREAT READS BEFORE YOU GO

Laura Catena, *Vino Argentino: An Insider's Guide to the Wines and Wine Country of Argentina* (Chronicle Books, 2010).

Ian Mount, *The Vineyard at the End of the World: Maverick Winemakers and the Rebirth of Malbec* (Norton, 2012).

Peter Richards, *The Wines of Chile* (Mitchell Beazley, 2006). The book is out of print but available as an e-book.

Greg de Villiers and Juan Vásquez, *Guide to the Wineries and Wines of Uruguay* (Editorial Planeta, 2012). Currently available only in Uruguay.

ONLINE RESOURCES

The trade association websites are great sources of information for visitors.

Wines of Argentina: www.winesofargentina.org

Wines of Brazil: www.winesofbrasil.com

Wines of Chile: www.winesofchile.org

Wines of Uruguay: www.winesofuruguay.com

Wine Sur: www.winesur.com. A great Argentinean wine site with articles, links to blogs, and lots of information about the country. Although it's targeted at the trade, there's ample information for anyone curious about Argentina's wine country.

Argentina National Tourism's very good general website also has a respectable wine section that provides regional information: www.argentina.travel/en/gourmet /the-wine-route.

IBRAVIN: www.ibravin.org.br. The official website of the Brazilian Wine Institute. Again, a lot of its information is intended for the trade, but the site is helpful to all who are curious, and it includes a consumer section.

Vale dos Vinhedos: www.valedosvinhedos.com.br. The website of Brazil's leading wine region has information, recommended wine routes, and listings for restaurants, food, and hotels. Worth exploring before you embark.

INAVI: www.inavi.com.uy. The official website of the Uruguayan Wine Institute. A website geared to the trade but still of interest to consumers. Currently only in Spanish.

Family Wineries of Uruguay "Wine Roads": www.uruguaywinetours.com. A one-stop shop for visiting family-run wineries in Uruguay. Great for setting up appointments.

Wine Hotels Collection: www.winehotelscollection.com. A global directory of wine country and winery hotels, wineries, vineyards, and restaurants with extensive sections on Argentina, Chile, and Uruguay.

Several of Chile's wine regions maintain their own websites:

www.colchaguavalley.cl

www.rutadelvinocurico.cl

www.maipoalto.com (in Spanish only)

www.casablancavalley.cl

WINERIES TO VISIT (AND WHY YOU SHOULD VISIT THEM)

Here I have assembled, by country, lists of wineries to visit. All of them make good wine (of course), but several of them offer other attractions as well: great architecture and design, an excellent restaurant, or a particularly good visitor experience. These attractions are listed below. Please refer to the individual winery profiles in each chapter for contact information and details about planning a visit; some wineries are open to visitors by appointment only.

ARGENTINA

Alta Vista, Luján de Cuyo (picnic grounds)

Andeluna Cellars, Uco Valley (design)

Belasco de Baquenado, Luján de Cuyo (aroma room)

Bodega Atamisque, Uco Valley (architecture, trout farm, golf)

Bodega Norton, Luján de Cuyo (restaurant)

Bodega Ruca Malén, Luján de Cuyo (restaurant)

Bodega Vistalba, Luján de Cuyo (restaurant)

Bodegas Escorihuela Gascón, Mendoza City (restaurant)

Callia, San Juan (design)

Catena Zapata, Luján de Cuyo (hospitality, architecture)

Colomé, Salta (architecture, art, museum, restaurant, hiking)

DiamAndes, Uco Valley (design)

Domingo Molina, Salta (artisan cheeses, view)

Familia Schroeder, Neuquén (museum, restaurant)

Finca las Nubes, Salta (view, interactivity, caves)

Graffigna, San Juan (museum, hospitality)

La Riojana, La Rioja (hospitality)

Melipal, Luján de Cuyo (restaurant)

Michel Torino, Salta (restaurant, hotel)

NQN, Neuquén (restaurant)

O. Fournier, Uco Valley (architecture, design, restaurant)

Piattelli Vineyards, Luján de Cuyo (hospitality)

Rutini, Luján de Cuyo (museum)

Salentein, Uco Valley (architecture, restaurant)

Terrazas de los Andes, Luján de Cuyo (design, restaurant)

BOLIVIA

Aranjuez, Tarija (gardens, design)

Campos de Solana, Tarija (hospitality, architecture)

Casa Grande, Tarija (restaurant, spa)

Kohlberg, Tarija (historic property, gardens)

La Concepción, Tarija (vista, history)

BRAZIL

Aurora, Vale dos Vinhedos (hospitality, tour)

Boscato, Serra Gaúcha (hospitality, tour)

Casa Valduga, Vale dos Vinhedos (restaurant, hospitality)

Cordilheira de Sant'Ana, Campanha Gaúcha (view, hospitality)

Dal Pizzol, Serra Gaúcha (picnic grounds, museum, self-guided vineyard tour)

Dezem, Paraná (design)

Don Giovanni, Serra Gaúcha (restaurant)

Don Guerino, Serra Gaúcha (architecture, tour)

Don Laurindo, Serra Gaúcha (hospitality)

Geisse, Serra Gaúcha (hospitality)

Miolo Winery, Vale dos Vinhedos (hospitality, tours)

Perini, Serra Gaúcha (hospitality)

Pizzato, Vale dos Vinhedos (prearranged group lunches)

Salton, Serra Gaúcha (hospitality, tours)

Sanjo, Santa Catarina (architecture)

Vinibrasil, São Francisco (view)

CHILE

Antiyal, Maipo (tour, biodynamic farming)

Casa Marín, San Antonio (tours, gallery, hospitality)

Casa Silva, Colchagua (restaurant, horse shows, horseback riding, polo)

Casas del Bosque, Casablanca (restaurant)

Concha y Toro Pirque Wine Center, Maipo (gardens, hospitality, wine bar, restaurant)

Cono Sur, Colchagua (bike trails on the property)

Cousiño Macul, Maipo (tour, hospitality)

Gillmore, Maule (restaurant, hotel)

Haras de Pirque, Maipo (equestrian activities)

J. Bouchon, Maule (mountain biking, hospitality, equestrian activities)

Lapostolle, Colchagua (restaurant, boutique hotel)

Loma Larga, Casablanca (equestrian activities)

Matetic, San Antonio (hotel, restaurant, organic farming)

Miguel Torres, Curicó (restaurant)

MontGras Properties, Colchagua (hospitality, interactivity)

Morandé, Casablanca (hospitality, tour, restaurant)

Santa Carolina, Maipo (historic property, hospitality)

Santa Rita, Maipo (restaurant, museum)

Tabalí, Limarí (ancient rock carvings)

Undurraga, Maipo (hospitality)

Veramonte, Casablanca (tour, hospitality)

Via Wines, Maule (hotel)

Vik, Colchagua (lodge, spa, horseback riding, mountain biking)

Viña Montes, Colchagua (hospitality, design, tours, restaurant)

Viu Manent, Colchagua (restaurant, equestrian activities)

URUGUAY

Alto de la Ballena, Maldonado (view)

Bertolini & Broglio, Paysandú (tour)

Bodega Garzón, Maldonado (design, architecture, olive oil production)

Bodegas Carrau, Canelones (historic property, hospitality)

Bodegas Castillo Viejo, Canelones (hospitality)

Bouza, Canelones (restaurant, antique car museum)

Familia Irurtia, Colonia (tour, historic cellar and vineyards)

Filgueira, Canelones (hospitality)

Finca Narbona, Colonia (tour, design, lodge, restaurant)

Juanicó, Canelones (tour, restaurant, hospitality)

Los Cerros de San Juan, Colonia (architecture, hospitality)

Marichal, Canelones (hospitality)

Viña Varela Zarranz, Canelones (historic property)

DINING SOUTH AMERICAN STYLE

The evolution in the vineyards and wineries of South America's wine country has not always been matched by progress in wine-country kitchens. Unquestionably, you can find world-class, inventive cuisine in many cities. Lima, for example, is home to one of the world's most respected celebrity chefs, Gastón Acurio, whose La Mar Cebichería Peruana has branches in Mexico City, Panama City, São Paulo, San Francisco, Chicago, and New York. The original restaurant of the Acurio Empire, Astrid y Gastón, has locations in Lima, Santiago, Buenos Aires, Madrid, Quito, Bogotá, Caracas, and Mexico City. In São Paulo, you can delight in the cutting-edge efforts led by Alex Atala, whose flagship restaurant, D.O.M., was ranked number 4 by the website The World's 50 Best Restaurants 2012 (www.theworlds50best.com). His more casual Dalva e Dito is also excellent, offering reinventions of traditional Brazilian foods. Helena Rizzo at Maní in São Paulo is another leading proponent of a reborn tradition.

Many chefs have embraced a more modern approach to their food, with an emphasis on locally produced ingredients. In Chile, Rodolfo Guzmán, of Boragó in Santiago, offers extraordinary cuisine that focuses on 100 percent Chilean ingredients, often cooked on hot stones and smoked with native woods as the Mapuche and Pehuenche Indians did centuries ago. In the Chilean wine country, you may be fortunate enough to dine at the hands of Pilar Rodríguez, whose food and wine studio at Viu Manent in the Colchagua Valley works closely with the entire wine industry. Pilar's mission is to celebrate the Chilean culinary bounty and pair it with her wines, which I can say from experience she does quite well. On the other side of the Andes, in Argentina, similar local-food and culturally connected approaches are shared by the "two Pablos" of Mendoza—Pablo del Río of the regionally diverse Siete Cocinas ("seven cuisines") and Pablo Ranea, the recently departed head chef of the very hip Azafrán. Both men are passionate about food, with del Río dedicated to showcasing the classic dishes of Argentina's seven regions in an up-to-date style, and Ranea adopting a modern approach that proves Argentine cuisine is no longer stalled out at steak and empanadas but is working its way toward a *nueva cocina Argentina*. Nadia O.F., in Mendoza City's suburb of Chacras de Coria, is owned and operated by Nadia Harón, wife of O. Fournier's chairman, José Manuel Ortega Fournier. The menu, which changes weekly, focuses on local ingredients and is created to pair with the wine offerings.

These stellar offerings are, however, the exceptions and not the rule, especially in wine country. South America is *muy tradicional* when it comes to food. Each country has an identity grounded in a handful of dishes. All of them share the empanada (meaning "wrapped in bread"), a stuffed bread or pastry baked or fried, depending on tradition and local culture. The Brazilian version is called *pastel*. It is in Argentina, however, that empanadas are king: Pablo del Río notes that there are seventeen regional versions.

Beyond empanadas, on the pampas (open grass plains) of Argentina, Brazil, and Uruguay, grilled beef rules. The Argentine *asado*, Brazilian *churrasco*, and Uruguayan *parrilla* are celebrations of beef, other grilled meats, and offal, accompanied by token salads. With the residents of the three countries consuming more than 140 pounds of beef per capita a year, vegetarians get short shrift. In Chile, too, people enjoy their meat: despite the country's vast coastline, seafood is very expensive, and most of the catch is destined for export. You'll still find amazing seafood in restaurants (including superb sea urchin), but the simple *milanesas* (breaded-meat cutlets) are the restaurant staple here—and also in Argentina, Bolivia, Brazil, Chile, Colombia, Paraguay, Peru, Uruguay, and Venezuela. Throw in pizzas and pasta in Argentina, pasta and roast chicken in southern Brazil, and varied regional protein and simple vegetable stews (*chupe* in Chile, *locro* in Argentina, and the heartier *feijoada*, a bean and meat stew, in Brazil) and you have South American everyday dining in a nutshell. The simplicity of these cuisines is at once a strength and a weakness, especially in wine country.

Most travelers to South American wine country start out wanting to dine native, devouring platters of empanadas and grilled meat, washed down by bottles of red wine (maybe there's a divine rationale for all that beef), and finishing with native grappas and *licors* served with *alfajores*, the cookie sandwiches filled with dulce de leche, or with Brazilian *quindim*, a rich coconut flan. This is great for a few days, but if you start hankering for variety in Bento Gonçalves, small-town Canelones, or the Mendoza wine country, good luck. Compared to the offerings of Adelaide or Healdsburg, South American cuisine is not super cosmopolitan. And that's (mostly) fine with the locals.

Happily, many of the best interpretations of local cuisine can be found at restaurants affiliated with wineries. In Argentina, Familia Zuccardi's Casa del Visitante, Francis Mallmann's 1884 (at Bodegas Escorihuela Gascón), and La Bourgogne at Carlos

Pulenta's Bodega Vistalba winery are classics. Uruguay boasts Finca Narbona's Finca y Granja in Colonia and a great restaurant at the Bouza Winery in Melilla (in Canelones). In Chile, you find Casas del Bosque's award-winning Tanino in Casablanca, overseen by chef Álvaro Larraguibel, and the very good Restaurante Viña Torres at the Miguel Torres winery in Curicó. Casa Valduga, in Brazil's Vale dos Vinhedos, is a standout, as is the dining room at the Spa Do Vinho, co-owned by the Miolo Winery across the street. Beyond those, and the restaurants noted above in Mendoza and Santiago, there are several wine-country restaurants worthy of attention.

ARGENTINA

In Mendoza, Tupungato Divino is part restaurant, part art gallery, part hotel, and a great stop in the Uco Valley. Another is Tunuyán's La Posada de Jamón, which specializes in (you guessed it) all things pork. If you find yourself in the Uco Valley in the evening, Ilo in downtown Tupungato is quite cool, with over one hundred different fish dishes and an ample wine list. Mendoza's Don Mario is a traditional, high-quality *asador*. In Patagonia, Malma Resto, NQN's restaurant, is a star.

BRAZIL

Bento Gonçalves is home base for almost all wine touring in Serra Gaúcha. The best dining is traditional: Canta Maria is the most famous example and quite good. For something a little more contemporary, try Sapore & Piacere near the church, whose chef, Márcia Dalla Chiesa, is very talented. In Campanha Gaúcha, my best meal by far was at Churrascaria Betemps in Bagé.

CHILE

In Casablanca, Bota Restaurante offers up an array of different local specialties with an international flair. In Romeral, close to Curicó, is the renowned Colo Colo Restaurante. This establishment is adored by locals and known for the country's best *plateadas* (a classic Chilean cut of beef), local cheeses, and *patitas de chancho* (pigs' trotters). Just outside Curicó is Cecinas Soler, a family affair considered *the* best place for sandwiches in Chile; it's been serving happy diners for more than thirty years. In San Antonio, La Juanita Restaurante serves traditional port-town seafood and is known for both its fried eel and its seafood soup.

URUGUAY

Montevideo's best *parrillada* is at La Casa Violeta. But I wouldn't rule out El Palenque, in Montevideo's bustling Mercado del Puerto. Also in Montevideo is Paninis, an Italian restaurant with a great wine cellar and pasta that many people consider the best in town. At Playa de José Ignacio, about twenty-five miles from Punta del Este in Maldonado, La Huella Restaurant is fantastic. The chef-owners, Martín Pittaluga and Guzmán Artaga-veytia, used to work with the Argentinean super-chef Francis Mallmann.

PERU

Most South American chefs will tell you that the epicenter for South American food is Peru. I've been told that spending the week at Lima's annual food festival, Mistura, which showcases Peruvian ingredients, traditional dishes, and culinary trends, is an absolute must. In 2011, around four hundred thousand people attended, including eminent global chefs such as Ferran Adrià (Spain), Dan Barber (United States), Heston Blumenthal (United Kingdom), Michel Bras (France), Alex Atala (Brazil), Rene Redzepi (Denmark), Yukio Hattori (Japan), Massimo Bottura (Italy), and of course the Peruvian celebrity chef Gastón Acurio. It is held at El Campo de Marte in early September. I'm waiting for my chance.

SUPER SOUTH AMERICAN SELECTIONS

BY VARIETY

TITILLATING TORRONTÉS

Bodegas Etchart "Gran Linaje"	Argentina
Colomé	Argentina
Crios de Susana Balbo	Argentina
Domingo Molina	Argentina
Finca Las Nubes	Argentina
Manos Negras	Argentina
Michel Torino "Finca La Primavera"	Argentina
Monteviejo "Festivo"	Argentina
O. Fournier "Urban Uco"	Argentina
Terrazas de los Andes Reserva	Argentina

SUPER SAUVIGNON BLANC

Casa Marín "Cipreses Vineyard"	Chile
Casas del Bosque "Pequeñas Producciones"	Chile
Casa Silva "Cool Coast"	Chile
Leyda "Single Vineyard Garuma"	Chile

Morandé "Edición Limitada"	Chile
O. Fournier "Alfa Centauri"	Chile
Pizzorno Reserva	Uruguay
Sanjo "Núbio"	Brazil
San Pedro "1865 Single Vineyard"	Chile
Santiago Queirolo "Intipalka" Sauvignon Blanc	Peru

ENCHANTING CHARDONNAY

Catena Zapata "White Stones" Adrianna Vineyard	Argentina
Clos des Fous "Locura 1"	Chile
Concha y Toro "Marques de Casa Concha" Chardonnay	Chile
Cordilheira de Sant'Ana Chardonnay	Brazil
Errázuriz "Wild Ferment"	Chile
Lapostolle "Cuvée Alexandre" Atalayas Vineyard	Chile
Luca Winery Chardonnay	Argentina
Maycas del Limarí "Quebrada Seca" Chardonnay	Chile
Tabalí "Talinay"	Chile
Viña Aquitania "SoldeSol"	Chile

RESPLENDENT ROSADO

Alto de la Ballena Rosado	Uruguay
Crios de Susana Balbo Rose of Malbec	Argentina
Doña Paula Los Cardos Malbec Rosado	Argentina
Dunamis "TOM"	Brazil
Falernia Syrah Rosado	Chile
Finca Narbona Tannat Rosado	Uruguay
Melipal Malbec Rosado	Argentina
Miguel Torres "Las Mulas" Rosado	Chile
Trapezio Malbec-Merlot Rosado	Argentina
Viña Progreso Syrah Rosado	Uruguay

PRIMO PINOT NOIR

Bodegas Carrau "Juan Carrau Reserva"	Uruguay
Casa Marín "Lo Abarca Hills"	Chile
Chacra "32"	Argentina
Cono Sur "20 Barrels"	Chile
Humberto Canale Reserva	Argentina
Kingston Family "Alazan"	Chile

Lidio Carraro "Dádivas"	Brazil
Marcelo Miras "UDWE"	Argentina
Veramonte "Ritual"	Chile
Viña Casablanca "Cefiro"	Chile

MEMORABLE MERLOT

Basso Monte Paschoal "Dedicato"	Brazil
Boscato Gran Reserva	Brazil
De Lucca Reserva	Uruguay
Dunamis	Brazil
Lapostolle "Cuvée Alexandre"	Chile
Miolo "Terroir"	Brazil
Pizzato "DNA99"	Brazil
Salton "Desejo"	Brazil
Santa Ema Reserve	Chile
Viña Cobos "Felino"	Argentina

MAGICAL MALBEC

Alta Vista "Alizarine"	Argentina
Belasco de Baquedano "Swinto"	Argentina
Doña Paula "Selección de la Bodega"	Argentina
Durigutti Familia Malbec	Argentina
Finca Sophenia Synthesis Malbec	Argentina
Goulart "Grand Vin"	Argentina
Luigi Bosca D.O.C.	Argentina
Monteviejo "Lindaflor"	Argentina
O. Fournier "Alfa Crux" Malbec	Argentina
Viu Manent	Chile

BONARDA AND CARIGNAN

Altos Las Hormigas "Colonia Las Liebres" Bonarda	Argentina
De Martino "Vigno" Carignan	Chile
Familia Zuccardi "Serie A" Bonarda	Argentina
Garage Wine Company Carignan "Vigno Lot #29"	Chile
Gillmore "Vigno" Carignan	Chile
Lamadrid Bonarda Reserva	Argentina
Morandé Loncomilla Edición Limitada Carignan	Chile
Nieto Senetiner Bonarda Reserva	Argentina
Odfjell "Vigno" Carignan	Chile
Trapiche Broquel Bonarda	Argentina

KILLER CARMENÈRE

Canepa "Reserva Privada"	Chile
Carmen "Nativa" Reserva	Chile
Casa Silva Gran Reserva "Los Lingues"	Chile
De Martino "Legado" Reserva	Chile
Odfjell "Orzada"	Chile
Tamaya "Winemaker's Selection"	Chile
Undurraga "Sibaris"	Chile
Viña Chocalán Reserva	Chile
Viña La Rosa "La Palma" Reserva	Chile
Viu Manent "El Incidente"	Chile

CABERNET—BUT NOT SAUVIGNON

Alto de la Ballena Cabernet Franc	Uruguay
Aurora "Pequenas Partilhas" Cabernet Franc	Brazil
Bodegas Castillo Viejo "Vieja Parcela" Cabernet Franc	Uruguay
Casa Valduga "Raizes" Cabernet Franc	Brazil
Dal Pizzol "Do Lugar" Cabernet Franc	Brazil
Doña Paula "Alluvia" Cabernet Franc	Argentina
Garage Wine Company "Lot 28" Cabernet Franc	Chile
Lagarde Cabernet Franc	Argentina
Lamadrid Reserva Cabernet Franc	Argentina
Leonardo Falcone Cabernet Franc	Uruguay

STELLAR SYRAH

Alto de la Ballena "Cetus"	Uruguay
Casas del Bosque "Pequeñas Producciones"	Chile
De Martino "Legado" Reserva	Chile
Falernia "Donna Maria"	Chile
Giménez Mendez "Alta Reserva"	Uruguay
Las Moras "Black Label"	Argentina
Matetic "Coralillo"	Chile
Mayu Reserva	Chile
Polkura	Chile
Ventisquero "Pangea"	Chile

CRAZY GOOD CABERNET SAUVIGNON

Benegas "Libertad"	Argentina
Bressia "Monteagrelo"	Argentina

Catena Zapata "Catena Alta"	Argentina
Concha y Toro "Don Melchor"	Chile
Cousiño Macul "Antiguas Reservas"	Chile
Koyle Reserva	Chile
Routhier & Darricarrère "Provincia de São Pedro"	Brazil
Santa Carolina "Reserva de Familia"	Chile
Undurraga "T.H." Pirque	Chile
Viña Cobos "Bramare Marchiori Vineyard"	Argentina

Antigua Bodega Stagnari "Prima Donna"	Uruguay
Bodegas Carrau "Amat"	Uruguay
Bouza "Parcela Única"	Uruguay
Don Guerino Reserva Tannat	Brazil
Finca Narbona "Luz de Luna"	Uruguay
H. Stagnari "Daymán"	Uruguay
Leonardo Falcone "Santa Cecilia"	Uruguay
Marichal Reserve Collection	Uruguay
Pisano "RPF" (Reserva Personal de la Familia)	Uruguay
Toscanini Reserva	Uruguay

BUBBLIES, BLENDS, AND OTHER BEST PICKS

SPECIAL SPARKLERS

Casa Valduga Cuvée "130" Brut	Brazil
Courmayeur Espumante Moscatel	Brazil
Cruzat "Cuvée Réserve" Extra Brut	Argentina
Don Giovanni "Ouro" Brut	Brazil
Geisse "Cave Geisse" Blanc de Blancs	Brazil
Geisse "Cave Geisse" Terroir Nature	Brazil
Marichal Blanc de Noirs Reserve Collection	Uruguay
Perini Moscatel	Brazil
Peterlongo "Presence" Brut	Brazil
Suzin Rosé	Brazil

AND NOW FOR SOMETHING COMPLETELY DIFFERENT

| Ariano Hermanos Muscat Ottonel | Uruguay |
| Aurora Riesling Italico | Brazil |

AND NOW FOR SOMETHING COMPLETELY
DIFFERENT *(continued)*

Bodega Garzón Albariño	Uruguay
Bodegas Carrau "1752" (Sauvignon Gris and Petit Manseng)	Uruguay
Dal Pizzol Touriga Nacional	Brazil
De Lucca Marsanne	Uruguay
Don Guerino Moscato Giallo	Brazil
Familia Irurtia "Km.0" Viognier	Uruguay
Filgueira Sauvignon Gris	Uruguay
Humberto Canale Riesling	Argentina
Juanicó Familia Deicas "Preludio" (50/50 Chardonnay and Viognier)	Uruguay
Kuhlmann AltoSama Demi Sec (sparkling wine)	Bolivia
Matetic "Corralillo" Riesling	Chile
Miguel Torres "Nectaria Botrytis" Riesling	Chile
Perini Marselan	Brazil
Pizzato Alicante Bouschet	Brazil
Pizzato Egiodola	Brazil
Tacama "Albilla d'Ica"	Peru
Viñedo de los Vientos "Alcyone" Licor de Tannat	Uruguay
Vista Alegre Tempranillo "Picasso"	Peru

BEST BOTTLES FOR BEEF

Antiyal	Chile
Bouza "Parcela Unica B6" Tannat	Uruguay
Casa Bianchi "Enzo Bianchi"	Argentina
Clos des Fous Cabernet Sauvignon	Chile
Kaiken "Mai"	Argentina
Los Boldos Gran Reserva Cabernet Sauvignon	Chile
Mendel "Unus"	Argentina
O. Fournier "Centauri Blend"	Chile
Trapiche "Single Vineyard Viña Domingo Sarmiento" Malbec	Argentina
Vik "Vik"	Chile

COMPLEX CORTES (BLENDS)

Antigua Bodega Stagnari "Mburucuyá"	Uruguay
Bouza "Monte Vide Eu"	Uruguay
Flaherty	Chile
Flechas de los Andes "Gran Corte"	Argentina

Lapostolle "Clos Apalta"	Chile
Rukumilla	Chile
Trapezio "Bo Bó"	Argentina
Viña Von Siebenthal "Carabantes"	Chile
Viñedo de Los Vientos "Eolo" Gran Reserva	Uruguay

BEST VALUES: WHITE

Alta Vista "Premium" Torrontés	Argentina
Andeluna "1300" Torrontés	Argentina
Aresti Sauvignon Blanc	Chile
Basso Moscatel	Brazil
Bisquertt "La Joya" Gewürztraminer	Chile
Bodegas Castillo Viejo "Reserva de la Familia" Sauvignon Blanc	Uruguay
Cono Sur "Organic" Sauvignon Blanc	Chile
Doña Paula Estate Chardonnay	Argentina
Falernia Pedro Ximénez Reserva	Chile
Familia Zuccardi "Santa Julia" Organic Torrontés	Argentina
Garibaldi Moscatel	Brazil
Giménez Méndez "100 Años" Sauvignon Blanc	Uruguay
Los Vascos Sauvignon Blanc	Chile
Marcelo Miras "UDWE" Semillon	Argentina
Michel Torino "Don David" Torrontés	Argentina
O. Fournier "Urban Uco" Sauvignon Blanc	Argentina
Pizzorno "Don Prospero" Sauvignon Blanc	Uruguay
RAR Viognier	Brazil
Tamaya Reserva Chardonnay	Chile
Zorzal Sauvignon Blanc	Argentina

BEST VALUES: RED

Achaval Ferrer Mendoza Malbec	Argentina
Bodega del Fin del Mundo Reserva Pinot Noir	Argentina
Bodega Norton "Privada" Malbec	Argentina
Bodega Vistalba "Tomero" Malbec	Argentina
Bodega y Cavas de Weinert "Carrascal" (Malbec, Cabernet Sauvignon, and Merlot)	Argentina
Concha y Toro "Casillero del Diablo" Carmenère	Chile
Emiliana "Novas Gran Reserva" Cabernet Sauvignon	Chile
Juanicó "Don Pascual" Reserve Marselan	Uruguay
La Concepción "Cepas de Altura" Syrah	Bolivia

Maquis "Lien" (Bordeaux blend)	Chile
MontGras Reserva Carmenère	Chile
NQN "Malma Finca la Papay" Malbec	Argentina
Pérez Cruz Reserva Estate Cabernet Sauvignon	Chile
Salton "Intenso" Teroldego	Brazil
TerraMater "Limited Reserve" Carmenère	Chile
Trivento "Amado Sur" Malbec	Argentina
Ventisquero Colchagua Reserve Carmenère	Chile
Veramonte "Primus" Cabernet Sauvignon	Chile
Viña Montes "Classic Series" Merlot	Chile
Zorzal Wines "Gran Terroir" Pinot Noir	Argentina

TWENTY WINES TO DRINK BEFORE YOU DIE

Achaval Ferrer "Finca Altamira"	Argentina
Almaviva	Chile
Bodega Aleanna "El Gran Enemigo"	Argentina
Bodega Noemía Patagonia Malbec	Argentina
Bressia "Conjuro" Malbec	Argentina
Catena Zapata "Adrianna Vineyard" Malbec	Argentina
Concha y Toro "Carmin de Peumo" Carmenère	Chile
Emiliana "Gê"	Chile
Enrique Foster "Firmado"	Argentina
Errázuriz "Kai" Carmenère	Chile
Juanicó Familia Deicas "Preludio"	Uruguay
Luca Winery "Nico"	Argentina
Mendel "Finca Remota" Malbec	Argentina
Neyen "Espíritu de Apalta"	Chile
Pisano "ArretXea"	Uruguay
Quebrada de Macúl "Domus Aurea" Cabernet Sauvignon	Chile
Seña	Chile
Viña Montes "Alpha M"	Chile
Viña Progreso "Sueños de Elisa"	Uruguay
Viña von Siebenthal "Toknar" Petit Verdot	Chile

DECODING SOUTH AMERICAN WINE LABELS

Regulations specifying what producers can claim on wine labels vary around the world. Here's a quick guide to the rules and conventions that apply to the labels of the major wine-producing countries of South America.

Chaptalization refers to the practice of adding sugar to the unfermented grape in order to increase the alcohol level of the finished wine. It is typically used in years when growing conditions inhibit grape ripening, so that the grapes have a lower sugar content and thus a lower potential alcohol level.

Varietal designation refers to the minimum percentage of a specific grape variety required for labeling the wine as a varietal wine.

Vintage date refers to the minimum percentage of wine required from the year specified on the label.

Appellation refers to the minimum percentage of wine required produced from grapes grown in the appellation named on the label.

	Chaptalization allowed?	Acidification allowed?	Varietal designation	Vintage date	Appellation
Argentina	No	Yes	85%	85%	85% for IP and 100 percent for DOC appellations and for the Uco Valley
Bolivia	—	—	85%*	90%*	100% for specific regions*
Brazil	Under approved circumstances	Under approved circumstances	75% for wines sold in Brazil; 85% for export wines	100%	85% for appellation; 100% for DO
Chile	Yes	Yes	75% for wines sold in Chile; 85% for export wines	75% for wines sold in Chile; 85% for export wine	
Colombia	—	—	—	—	—
Ecuador	Yes	Yes, if needed	†	†	†
Paraguay	—	—	—	—	—
Peru	No‡	Yes	75%	75%	None
Uruguay	For export wines only	Yes	85%	None	None
Venezuela	—	—	—	—	—

*Not formally regulated, but most Bolivian wineries adhere to global standards for varietal designations, vintage, and appellation.
†No standards that are legally binding; the wineries voluntarily follow international standards.
‡Although chaptalization is not allowed, the use of concentrated grape must is permitted.

SOURCES

Anderson, Kym, and Signe Nelgen. *Global Wine Markets, 1961 to 2009: A Statistical Compendium*. Adelaide: University of Adelaide Press, 2009.

Atkin, Tim. "Carignan: Chile Returns to Its Roots." *Tim Atkin MW* (blog), www.timatkin.com. July 20, 2012.

Bigongiari, Diego, ed. *Viñas, bodegas y vinos de América del Sur/Vineyards, Wineries and Wines of South America 2005*. Argentina: Austral Spectator, 2004. (In Spanish and English.)

Burgos, Christian. "The Brazilian Thirst for Wine." *Meininger's Wine Business International*. August 30, 2012. www.wine-business-international.com/156-bWVtb2lyX2lkPTQ1NSZtZW51ZV9jYXRfaWWQ9--en-magazine-magazine_detail.html.

Catena, Laura. *Vino Argentino: An Insider's Guide to the Wines and Wine Country of Argentina*. San Francisco: Chronicle Books, 2010.

Caucasia Wine Thinking. *Exportaciones de la industria vitivinicola Argentina: Enero a diciembre*. Wines of Argentina, 2012, www.winesofargentina.org.

Decanter magazine. Various issues. www.decanter com.

DeSimone, Mike, and Jeff Jenssen. *Wines of the Southern Hemisphere*. New York: Sterling Epicure, 2012.

Foix, Augusto/Juíaterruños. *Wineries Resto Hotels: Food, Wine and Travel Guide*. 5th ed. Argentina: Grupo Terruño SA, 2011. (In Spanish and English.)

Frank, Jade Frank. "Chile's Elqui Valley: A Zen Experience in 'Travel Therapy.'" *GoNomad*, n.d., www.gonomad.com/3006-chile-s-elqui-valley-a-zen-experience-in-travel-therapy#ixzz2qbNZa3Mi. Accessed January 17, 2014.

Marino, Javier, ed. *Wineries and Wines of Argentina: International Yearbook 2011*. Area del Vino. Mendoza, Argentina, 2011.

Rice, Prudence. *Vintage Moquegua: History, Wine and Archaeology on a Colonial Peruvian Periphery*. Austin: University of Texas Press, 2011.

Richards, Peter. *The Wines of Chile*. London: Octopus, 2006.

Robinson, Jancis. *The Oxford Companion to Wine*. 3rd ed. New York: Oxford University Press, 2006.

Robinson, Jancis, Julia Harding, and José Vouillamoz. *Wine Grapes*. New York: Harper Collins, 2012.

Rolland, Michel, and Enrique Chrabolowsky. *Wines of Argentina*. 3rd ed. Argentina: Mirol SA, 2008.

Stein, Steve. "Culture and Identity: Transformations in Argentine Wine, 1880–2011." Unpublished paper. http://estructuraehistoria.unizar.es/gihea/documents/Stein.pdf. Accessed January 15, 2014.

Tapia, Patricio. *Descorchadas 2012: Guía de vinos de Argentina*. Buenos Aires: Simposium, 2012. (In Spanish.)

———, ed. *El valle de Casablanca*. Chile: PMC Communication, 2010.

Waldin, Monty. *Wines of South America*. London: Mitchell Beazley, 2003.

Wine and Spirits magazine. Various issues. www.wineandspirits.com.

Wine Spectator. Various issues. www.winespectator.com.

INDEX

Page references given in *italics* indicate illustrations or material contained in their captions.

Canary Island immigrants, 215–16

Candiota (Brazil), *179*, 190

Canelón Chico (Uruguay), 227, 228

Canelones (Uruguay), 15, 19–20, 25, 33–34, *207*, 208, 215–16, 222, 223, 227

Canepa, 122, 124, 132; "Genovino," 132; "Magnificum," 132; Reserva Privada, 268

Canepa sisters (Edda, Antonieta, and Gilda), 164

Canepa Vacarreza, Giuseppe, 132

Cañete (Chile), 150

Cañete-Lima (Peru), 38

Canta María (restaurant), 263

Cantón, Eduardo, 223

Cap Artigas (Uruguay), 216

Carabantes, Francisco de, 7

Caracas (Venezuela), 261

Cardozo, Alejandro, 198

Carignan, 25–26; Argentina, 85, 105–6; Chile, 25–26, 111, 126, *128*, 132, 142, 152, 157, 166; recommended selections, 267

Carlos Pulenta Wines, 71–72

Carmelo (Uruguay), 215, 223, 226

Carmen, 122, 129, 132–33, 162; "Gold Reserve," 132; "Nativa" Reserva, 268

Carmenère, 25, 26–27; Brazil, 194; Chile, 26–27, 30, 111, 113, 122, 123, 124, 129, 133, 134, 135, 141, 144, 152, 153, 155, 156, 165, 175, 177; recommended selections, 268

Carolina Wine Brands, 133, 156, 162, 170. *See also* Ochagavía; Santa Carolina; Viña Casablanca

Carora (Venezuela), 243

Carraro, Lidio, 11, 189–90, 200

Carraro family, 181, 200

Carrau. *See* Bodegas Carrau

Carrau, Francisco, 11, 210–11

Carrau, Juan, 211, 220

Carrau Vehils, Francisco, 220

Carrodilla (Argentina), 84

Cartagena (Colombia), 236, 237

Carta Vieja, 111

Casa Bianchi, 60, 76; "Enzo," 76, 270

Casablanca (Chile), 9, 14, 19, 31, 33, 109, 118–19, 133, 146, 147, 149, 152, 154, 159, 160, 162, 164–65, 168, 169, 170, 172, 263

Casa del Visitante (restaurant), 262

Casa Grajales, 237

Casa Grande, 235, 244–45, 257; "Osadía," 245; Reserva Trivarietal, 245

Casa Marín, 15, 18, 19, 21, 120, 133, 258; "Cipreses Vineyard" Sauvignon Blanc, 265; "Lo Abarca Hills" Pinot Noir, 133, 266

Casa Nieto: Buenos Aires, Argentina, 97; São Paulo, Brazil, 97

Casa Nova (Brazil), 200

Casa Real *(singani)*, 244

Casarena, 44

Casa Rivas, 122, 125, 134, 143; Gran Reserva Carmenère, 134

Casas del Bosque, 19, 33, 119, 134, 258, 263; Gran Estate Selection, 134; "Pequeñas Producciones," 265, 268

Casas del Toqui, 123, 134–35; "Gran Toqui," 134

Casas de Pirque, 160–61. *See also* Santa Alicia

Casa Silva, 27, 124, 135, 183, 198, 258; "Cool Coast," 265; Gran Reserva "Los Lingues," 268

Casa Valduga, 11, 22, 24, 180, 181–82, 189, *191*, 193, 257, 263; Cuvée "130" Brut, 269; "Raizes" Cabernet Franc, 24, 268; "Storia," 193

Casell, Laura, 218

Castellano family, 244

Castel Pujol, 220

Castillo Viejo, 20. *See* Bodegas Castillo Viejo

Catamarca (Argentina), 46, 51, 75, 93

Catena, Adrianna, 66

Catena, Ernesto, 73, 84, 103

Catena, Jorge, 10

Catena, Laura, 23, 33, 77, 84, 92, 93, 255

Catena, Nicolás, 10, 45, 72–73, 76–77, 84, 102, 183

Catena, Tikal, 103

Catena family, 72, 76–77

Catena Family of Wines. *See* Bodegas Escorihuela Gascón; Bodegas Esmeralda; Catena Zapata; Ernesto Catena Vineyards; Tikal Winery

Catena Zapata, 14, 25, 29, 44, 48, 56, 76–77, 93, 256; "Adrianna Vineyard" Malbec, 272; "Catena Alta," 269; "White Stones," 14, 266

Catholic missionaries, 7, 153, 190, 214–15, 246

Catholic University of Chile (Santiago), 113

Catrala, 119, 135; Grand Reserve Limited Edition Pinot Noir, 135

Caucete (Argentina), 53

Cauquenes (Chile), 26, 126, 132, 142

Cavalleri, 189, 193–94; Merlot Reserva, 194

Caxias do Sul (Brazil), 189

Cecchin family, 85

Cecinas Soler (restaurant), 263

Cedron, Juan, 7

Celeste, Gabriela, 101–2

CENAVIT (Bolivian National Viticultural Center), 234–35

Central Valley (Chile), 111; about, 120, *121 map*;

de Superunda," 153; "Estelado" Santa Digna, 36; Estelado sparkling wine, 153; "Las Mulas" Rosado, 266; "Nectaria Botrytis" Riesling, 270; Reserva de Pueblo, 153

milanesas, 262

Milantino, 183

Millahue (Chile), 169

Minas Gerais (Brazil), 186

Miolo, Adriano, 11, 183–84, 200

Miolo family, 181, 190, 200

Miolo Wine Group, 180, 181, 200

Miolo Winery, 11, 30, 35, 182, 189, 200–201, 257, 263; "Castas Portugeisas," 35; "Lot 43," 184, 201; Reserva, 201; Seleção, 201; "Terroir" Merlot, 30, 201, 267

Miraflores (Chile), 149

Miranda, Pilar, 142

Miras, Marcelo and Pablo, 94

Misiones de Rengo, 123, 125, 153; Gran Reserva Cuvée Carmenère, 153

Mission grape. *See* Pais

Mistura Food Festival, 264

Mizque (Bolivia), 8, 233

Moët, Claude, 78

Moët & Chandon, 10–11, 46, 78, 181

Moët Hennessy, 103

Molina (Chile), 139, 160, 173

Molina, Dora and Carlos, 245

Molinos (Argentina), 50

Molinos Río de la Plata, 96–97

Mollar, 6, 27, 36, 37

Monastrell, 153

Monceau, Jean du, 67

Mondavi, Robert, 10

Montalembert, Jacques Louis de, 70

Monte Belo do Sul (Brazil), 189, 194

Monte Bérico (Brazil), 192

Monte Patria (Chile), 117

Montepulciano: Brazil, 199

Montes, Aurelio, 90, 91, 113, 146, 172

Montes Wines. *See* Viña Montes

Montevideo (Uruguay), 208, 263–64

Monteviejo, 59, 96; "Festivo," 265; "Lindaflor," 96, 267

Montgolfier, Ghislain de, 169–70

MontGras, 124, 154, 155, 258; Reserva Carmenère, 154, 272

MontGras Properties, 154. *See also* MontGras; Ninquén

Moquegua (Peru), 242

Morandé, 26, 27, 119, 154, 258; "Edición Limitada"

Carignan, 267; "Edición Limitada" Sauvignon Blanc, 266; "House of Morandé," 154

Morandé, Pablo, 9, 19, 118, 154

Moreno, Yerko, 135

Morro da Fumaça (Brazil), 187

Moscatel (Brazil), 37, 182, 184, 194, 215

Moscato Bianco, 16

Moscato Giallo, 16; Brazil, 16, 196, 206

Moscato R2, 16

Mossman Knapp, Derek, 111, 126, 142

Mounier, José Luis, 88

Mount, Ian, 255

Mourvèdre, 249

Movimiento de Viñateros Independientes (MOVI), 111, 126, 139, 142, 153, 157, 161, 169, 175

mugrón (propagation technique), 111

Mulchén (Chile), 127, 139

Mumm Argentina, 53, 59, 99

Muñoz, Gonzalo, 158

Murray, Douglas, 90

Muscat à Petits Grains, 16; Uruguay, 229

Muscat Bailey A, 16

Muscat, 6, 15–16, 38; Argentina, 16; Brazil, 16, 184, 188, 189, 192, 202; Chile, 16, 110, 116, 127; Uruguay, 16. *See also* Moscatel; Moscato Giallo; Rosado Moscato

Muscat of Alexandria, 15–16, 37; Bolivia, 233, 235, 246; Chile, 127

Muscat of Hamburg (Black Muscat), 16

Muscat Ottonel, 16, 213

Mutio family, 228

Nadia O.F. (restaurant), 262

Nancagua (Chile), 123

Napa Valley (CA, USA), 10

Narbona. *See* Finca Narbona

Narbona, Juan de, 223

Narváez, Raúl, 145

National Agricultural Research Institute (Argentina), 73

National Distillers (Brazil), 190

National University of Cuyo, 46, 73

Nativa Eco Wines, 122, 155; Gran Reserva Cabernet Sauvignon, 155

Navarrete, Viviana, 114

Nazário dos Santos, Antonio, 203

Nazca Valley (Peru), 242

Nebbiolo: Argentina, 105–6; Brazil, 199; Chile, 131; Uruguay, 230

Negra Criolla, 233, 235, 246. *See also* Pais

Negramoll, 27, 36